T0324805

COMPUTATION, PROOF, MACHINE

Mathematics Enters a New Age

Computation is revolutionizing our world, even the inner world of the "pure" mathematician. Mathematical methods – especially the notion of proof – that have their roots in classical antiquity have seen a radical transformation since the 1970s, as successive advances have challenged the priority of reason over computation.

Like many revolutions, this one comes from within. Computation, calculation, algorithms – all have played an important role in mathematical progress from the beginning – but behind the scenes, their contribution obscured in the enduring mathematical literature. To understand the future of mathematics, this fascinating book returns to its past, tracing the hidden history that follows the thread of computation. Along the way it invites us to reconsider the dialogue between mathematics and the natural sciences, as well as the relationship between mathematics and computer science. It also sheds new light on philosophical concepts, such as the notions of analytic and synthetic judgment. Finally, it brings us to the brink of the new age, in which machine intelligence offers new ways of solving mathematical problems previously inaccessible.

This book is the 2007 Winner of the Grand Prix de Philosophie de l'Académie Française.

Mathematician, logician, and computer scientist **Gilles Dowek** is currently a researcher at the French Institute for Research in Computer Science and Automation (INRIA). He is a member of the scientific board of the Société informatique de France and of CERNA. He has also been a consultant with the National Institute of Aerospace, a NASA-affiliated lab. He is the recipient of the French Mathematical Society's Grand Prix d'Alembert des Lycéens for his popular science work.

Pierre Guillot is a lecturer in mathematics at the University of Strasbourg's Institute of Advanced Mathematical Research (IRMA).

Marion Roman is a translator based in France.

Computation, Proof, Machine

MATHEMATICS ENTERS A NEW AGE

Gilles Dowek

*French Institute for Research in Computer Science
and Automation (INRIA)*

Translated from the French by

Pierre Guillot *and* **Marion Roman**

CAMBRIDGE
UNIVERSITY PRESS

CAMBRIDGE
UNIVERSITY PRESS

University Printing House, Cambridge CB2 8BS, United Kingdom

One Liberty Plaza, 20th Floor, New York, NY 10006, USA

477 Williamstown Road, Port Melbourne, VIC 3207, Australia

4843/24, 2nd Floor, Ansari Road, Daryaganj, Delhi - 110002, India

79 Anson Road, #06-04/06, Singapore 079906

Cambridge University Press is part of the University of Cambridge.

It furthers the University's mission by disseminating knowledge in the pursuit of education, learning and research at the highest international levels of excellence.

www.cambridge.org
Information on this title: www.cambridge.org/9780521118019

Original edition:
Les Métamorphoses du calcul
© Editions le Pommier – Paris, 2007
English translation © Cambridge University Press 2015

First published as *Les Métamorphoses du calcul* 2007
First English edition 2015

A catalogue record for this publication is available from the British Library

Library of Congress Cataloging in Publication data
Dowek, Gilles.
Computation, proof, machine : mathematics enters a new age / Gilles Dowek, French Institute for Research in Computer Science and Automation (INRIA), Pierre Guillot, Université de Strasbourg, Marion Roman.
pages cm
Includes bibliographical references and index.
ISBN 978-0-521-11801-9 (hardback) – ISBN 978-0-521-13377-7 (pbk.)
1. Mathematics – History – 20th century.   2. Mathematics – History.
3. Mathematics, Ancient.   I. Guillot, Pierre, 1978–   II. Roman, Marion.
III. Title.   IV. Title: Mathematics enters a new age.
QA26.D69   2015
510.9–dc23   2015004116

ISBN 978-0-521-11801-9 Hardback
ISBN 978-0-521-13377-7 Paperback

To Gérard Huet

Contents

Introduction: In Which Mathematics Sets Out to
Conquer New Territories   *page* 1

Part One: Ancient Origins

1   The Prehistory of Mathematics and the Greek
    Resolution   5
2   Two Thousand Years of Computation   15

Part Two: The Age of Reason

3   Predicate Logic   31
4   From the Decision Problem to Church's Theorem   44
5   Church's Thesis   55
6   Lambda Calculus, or an Attempt to Reinstate
    Computation in the Realm of Mathematics   69
7   Constructivity   73
8   Constructive Proofs and Algorithms   82

Part Three: Crisis of the Axiomatic Method

9   Intuitionistic Type Theory   89
10 Automated Theorem Proving   96
11 Proof Checking   105
12 News from the Field   111
13 Instruments   124

14   The End of Axioms?   134
      Conclusion: As We Near the End of This
      Mathematical Voyage...   136

Biographical Landmarks   139
Bibliography   149

Introduction: In Which Mathematics Sets Out to Conquer New Territories

IT'S BEEN said again and again: the century that just ended was the true golden age of mathematics. Mathematics evolved more in the twentieth century than in all previous centuries put together. Yet the century just begun may well prove exceptional for mathematics, too: the signs seem to indicate that, in the coming decades, mathematics will undergo as many metamorphoses as in the twentieth century – if not more. The revolution has already begun. From the early seventies onward, the mathematical method has been transforming at its core: the notion of proof. The driving force of this transformation is the return of an old, yet somewhat underrated mathematical concept: that of computing.

The idea that computing might be the key to a revolution may seem paradoxical. Algorithms that allow us, among other things, to perform sums and products are already recognized as a basic part of mathematical knowledge; as for the actual calculations, they are seen as rather boring tasks of limited creative interest. Mathematicians themselves tend to be prejudiced against computing – René Thom said: "A great deal of my assertions are the product of sheer speculation; you may well call them reveries. I accept this qualification. ... At a time when so many scientists around the world are computing, should we not encourage those of them who can to dream?" Making computing food for dreams does seem a bit of a challenge.

Unfortunately, this prejudice against computing is ingrained in the very definition of mathematical proof. Indeed, since Euclid, a proof has been defined as reasoning built on axioms and inference rules. But are mathematical problems always solved using a reasoning process? Hasn't the practice of mathematics shown, on the contrary, that solving a problem requires the subtle arrangement of reasoning stages and computing stages? By confining itself to reasoning, the axiomatic method may offer only a limited vision of mathematics. Indeed, the axiomatic method has reached a crisis, with recent mathematical advances, not all related to one another, gradually challenging the primacy of reasoning over computing and suggesting a more balanced approach in which these two activities play complementary roles.

This revolution, which invites us to rethink the relationship between reasoning and computing, also induces us to rethink the dialogue between mathematics and natural sciences such as physics and biology. It thus sheds new light on the age-old question of mathematics's puzzling effectiveness in those fields, as well as on the more recent debate about the logical form of natural theories. It prompts us to reconsider certain philosophical concepts such as analytic and synthetic judgement. It also makes us reflect upon the links between mathematics and computer science and upon the singularity of mathematics, which appears to be the only science where no tools are necessary.

Finally, and most interestingly, this revolution holds the promise of new ways of solving mathematical problems. These new methods will shake off the shackles imposed by past technologies that have placed arbitrary limits on the lengths of proofs. Mathematics may well be setting off to conquer new, as yet inaccessible territories.

Of course, the crisis of the axiomatic method did not come out of the blue. It had been heralded, from the first half of the twentieth century, by many signs, the most striking being two new theories that, without altogether questioning the axiomatic method, helped to reinstate computing in the mathematical edifice, namely the theory of computability and the theory of constructivity. We will therefore trace the history of these two ideas before delving into the crisis. However, let us first head for remote antiquity, where we will seek the roots of the very notion of computing and explore the "invention" of mathematics by the ancient Greeks.

PART ONE

Ancient Origins

CHAPTER ONE

The Prehistory of Mathematics and the Greek Resolution

THE ORIGIN OF MATHEMATICS is usually placed in Greece, in the fifth century B.C., when its two main branches were founded: arithmetic by Pythagoras and geometry by Thales and Anaximander. These developments were, of course, major breakthroughs in the history of this science. However, it does not go far enough back to say that mathematics has its source in antiquity. Its roots go deeper into the past, to an important period that laid the groundwork for the Ancient Greeks and that we might call the "prehistory" of mathematics. People did not wait until the fifth century to tackle mathematical problems – especially the concrete problems they faced every day.

ACCOUNTANTS AND LAND SURVEYORS

A tablet found in Mesopotamia and dating back to 2500 B.C. carries one of the oldest traces of mathematical activity. It records the solution to a mathematical problem that can be stated as follows: if a barn holds 1,152,000 measures of grain, and you have a barn's worth of grain, how many people can you give seven measures of grain to? Unsurprisingly, the result reached is 164,571 – a number obtained by dividing 1,152,000 by seven – which proves that Mesopotamian accountants knew how to do division long before arithmetic was born. It is even likely (although it is hard to know anything for certain in that field) that writing was invented in order to keep account books and that, therefore, numbers were invented before letters. Though it may be hard to stomach, we probably owe our whole written culture to a very unglamorous activity: accounting.

Mesopotamian and Egyptian accountants not only knew how to multiply and divide, but had also mastered many other mathematical operations – they were able to solve quadratic equations, for instance. As for land surveyors, they knew how to measure the areas of rectangles, triangles, and circles.

THE IRRUPTION OF THE INFINITE

The techniques worked out by accountants and surveyors constitute the prehistory of arithmetic and geometry; as for the history of mathematics, it is considered to have begun in Ancient Greece, in the fifth century B.C. Why was this specific period chosen? What happened then that was so important? In order to answer this question, let us look at a problem solved by one of Pythagoras's disciples (whose name has been forgotten along the way). He was asked to find an isosceles right triangle whose three sides each measured a whole number of a unit – say, meters. Because the triangle is isosceles, its two short sides are the same length, which we will call x. The length of the hypotenuse (i.e. the triangle's longest side) we will call y. Because the triangle is right, y^2, according to Pythagoras's theorem, equals $x^2 + x^2$. To look for our desired triangle, then, let's try all possible combinations where y and x are less than 5:

x	y	$2 \times x^2$	y^2
1	1	2	1
1	2	2	4
1	3	2	9
1	4	2	16
2	1	8	1
2	2	8	4
2	3	8	9
2	4	8	16
3	1	18	1
3	2	18	4
3	3	18	9
3	4	18	16
4	1	32	1
4	2	32	4
4	3	32	9

In all these cases, the number $2 \times x^2$ is different from y^2. We could carry on searching, moving on to larger numbers. In all likelihood, Pythagoras's followers kept looking for the key to this problem for a long time, in vain, until they eventually became convinced that no such triangle existed. How did they manage to reach this conclusion, namely that the problem could not be solved? Not by trying out each and every pair of numbers one after the other, for there are infinitely many such pairs. Even if you tried out all possible pairs up to one thousand, or even up to one million, and found none that worked, you still could not state with any certainty that the problem has no solution – a solution might lie beyond one million.

Let's try to reconstruct the thought process that may have led the Pythagoreans to this conclusion.

First, when looking for a solution, we can restrict our attention to pairs in which at least one of the numbers x and y is odd. To see why, observe that if the pair $x = 202$ and $y = 214$, for example, were a solution, then, by dividing each number by two, we would find another solution, $x = 101$ and $y = 107$, where at least one of the numbers is odd. More generally, if you were to pick any solution and divide it by two, repeatedly if necessary, you would eventually come to another solution in which at least one of the numbers is odd. So, if the problem has any solution, there is necessarily a solution in which either x or y is an odd number.

Now, let's divide all pairs of numbers into four sets:

- pairs in which both numbers are odd;
- pairs in which the first number is even and the second number is odd;
- pairs in which the first number is odd and the second number is even;
- pairs in which both numbers are even.

We can now give four separate arguments to show that none of these sets holds a solution in which at least one of the numbers x and y is odd. As a result, the problem cannot be solved.

Begin with the first set: it cannot contain a solution in which one of the numbers x and y is odd, because if y is an odd number, then so is y^2; as a consequence, y^2 cannot equal $2 \times x^2$, which is necessarily an even number. This argument also rules out the second set, in which x is even and y odd. Obviously, the fourth set must also be ruled out, because by definition it cannot contain a pair where at least one number is odd. Which leaves us with the third set. In this case, x is odd and y is even, so that the number obtained by halving $2 \times x^2$ is odd, whereas half of y^2 is even – these two numbers cannot be equal.

The conclusion of this reasoning, namely that a square cannot equal twice another square, was reached by the Pythagoreans more than twenty-five centuries ago and still plays an important part in contemporary mathematics. It shows that, when you draw a right isosceles triangle whose short side is one meter long, the length of the hypotenuse measured in meters is a number (slightly greater than 1.414) that cannot be obtained by dividing x and y, two natural numbers, by each other. Geometry thus conjures up numbers that cannot be derived from integers using the four operations – addition, subtraction, multiplication, and division.

Many centuries later, this precedent inspired mathematicians to construct new numbers, called "real numbers." The Pythagoreans, however, did not go quite so far: they were not ready to give up what they regarded as the essential value of natural numbers. Their discovery felt to them more like a disaster than an opportunity.

Yet the Pythagorean problem was revolutionary not only because of its effects, but also because of how it is framed and how it was solved. To begin with, the Pythagorean problem is much more abstract than the question found on the Mesopotamian tablet, where 1,152,000 measures of grain were divided by 7 measures. Whereas the Mesopotamian question deals with measures of grain, the Pythagorean problem deals with numbers and nothing more. Similarly, the geometric form of the Pythagorean problem does not concern triangular fields but abstract triangles. Moving from a number of measures of grain to a number, from a triangular field to a triangle, may seem a trifle, but abstraction is actually a step of considerable importance. A field cannot measure more than a few kilometers. If the problem involved an actual triangular field, it would suffice, in order to solve it, to try every solution in which x and y are less than 10,000. But, unlike a triangular field, an abstract triangle can easily measure a million units, or a billion, or any magnitude.

Clearly, a rift had opened between mathematical objects, which are abstract, and concrete, natural objects – and this exists even when the mathematical objects have been abstracted from the concrete ones. It is this rift that was the big breakthrough of the fifth century B.C.

The growing distance between mathematical objects and natural ones led some people to think that mathematics was not fit to describe natural objects. This idea dominated until the seventeenth century – Galileo's day – when it was refuted by advances in mathematical physics. Yet it persists today in those views that deny mathematics any relevance in the fields of social sciences – as when Marina Yaguello argues that the role of mathematics in linguistics is to "cover up its 'social' (hence fundamentally inexact) science with complex formulae."

This change in the nature of the objects under study – which, since the fifth century B.C., have been geometric figures and numbers not necessarily related to concrete objects – triggered a revolution in the method used to solve mathematical problems. Once again, let's compare the methods used by the Mesopotamians and those used by the Pythagoreans. The tablet shows that Mesopotamians solved problems by performing computations – to answer the question about grain, they did a simple division. When it comes to the Pythagoreans' problem, however, reasoning is necessary.

In order to do a division, all you have to do is apply an algorithm taught in primary school, of which the Mesopotamians knew equivalents. By contrast, when developing their thought process, the Pythagoreans could not lean on any algorithm – no algorithm recommends that you group the pairs into four sets. To come up with this idea, the Pythagoreans had to use their imaginations. Maybe one of Pythagoras's followers understood that the number y could not be odd and then, a few weeks or a few months later, another disciple helped make headway by discovering that x could not be an odd number either. Perhaps it was months or even years before another Pythagorean made the next big advance. When a Mesopotamian tackled a division, he knew he was going to achieve a result. He could even gauge beforehand how long the operation would take him. A Pythagorean tackling an arithmetic problem had no means of knowing how long it would be before he found the line of reasoning that would enable him to solve the problem – or even if he ever would.

Students often complain that mathematics is a tough subject, and they're right: it is a subject that requires imagination; there is no systematic method for solving problems. Mathematics is even more difficult for professional mathematicians – some problems have remained unsolved for decades, sometimes centuries. When trying to solve a math problem, there is nothing unusual about drawing a blank. Professional mathematicians often stay stumped too, sometimes for years, before they have a breakthrough. By contrast, no one dries up over a division problem – one simply commits the division algorithm to memory and applies it.

How did the change in the nature of mathematical objects bring about this methodological change? In other words, how did abstraction lead mathematicians to drop calculation in favor of the reasoning that so characterizes Ancient Greek mathematics? Why couldn't the Pythagorean problem be solved by simple calculation? Think back, once more, to the Mesopotamian question. It deals with a specific object (a grain-filled barn) of known size. In the Pythagorean problem, the size of the triangle is not known – indeed, that's the whole problem. So the Pythagorean problem does not involve a specific triangle but, potentially, all possible triangles. In fact, because there is no limit to the size a triangle might reach, the problem concerns an infinity of triangles simultaneously. The change in the nature of the objects being studied is thus accompanied by the irruption of the infinite into mathematics. It was this irruption that made a methodological change necessary and required reasoning to be substituted for computing. For, if the problem concerned a finite number of triangles – for example, all triangles whose sides measure less than 10,000 metres – we could still resort to calculation. Trying out every possible pair of whole numbers up to 10,000 would doubtless be tedious without the aid of a machine, but it is nonetheless systematic and would settle the finite problem. As we've observed, though, it would be futile against the infinite.

This is why the transition from computing to reasoning, in the fifth century B.C. in Greece, is regarded as the true advent of mathematics.

THE FIRST REASONING RULES: PHILOSOPHERS
AND MATHEMATICIANS

One crucial question remains: what is reasoning? Knowing that all squirrels are rodents, that all rodents are mammals, that all mammals are vertebrates, and that all vertebrates are animals, we can infer that all squirrels are animals.

One reasoning process – among others – enabling us to reach this conclusion consists in deducing, successively, that all squirrels are mammals, then that all squirrels are vertebrates, and finally that all squirrels are animals. Although this process is extremely simple, its structure is not fundamentally different from that of mathematical reasoning. In both cases, the thought process is made up of a series of propositions, each of which follows logically from the previous one through the application of an "inference rule." Here we used the same rule three times in a row: "if all Y are X and all Z are Y, then all Z are X."

The Greek philosophers were the first to compile a list of these inference rules that enable new propositions to be deduced from those already established and hence allow reasoning processes to make headway. For example, we have Aristotle to thank for the aforementioned rule. Indeed, Aristotle set up a list of rules that he called *syllogisms*. Syllogisms can take on various forms. Some follow the "all Y are X" pattern, others fall into the "some Y are X" category. Thus, knowing that all Y are X, and that some Z are Y, we can infer that some Z are X.

Aristotle was not the only ancient philosopher to take an interest in inference rules. In the third century B.C., the Stoics laid out other such rules. One rule allows the proposition B to be deduced from the propositions "if A then B" and A.

These two attempts to catalog inference rules occurred contemporaneously with the development of Greek arithmetic and geometry, after the revolutionary methodological switch from computing to reasoning. It would have made sense for Greek mathematicians to use the logic of Aristotle or that of the Stoics to support their reasoning. In order to prove that a square cannot be twice another square, for instance, they might have resorted to a series of syllogisms. Strangely enough, they didn't, even though Greek philosophers and mathematicians clearly shared a common project. In the third century B.C., Euclid wrote a treatise in which he synthesized all the geometric knowledge of his time. He used deductive reasoning to prove his every point, yet, somewhat surprisingly, he never once called upon the logic of Aristotle or of the Stoics.

Several explanations can be put forward to account for this fact. The most likely hypothesis is that mathematicians rejected the philosophers' logic because it was too coarse. The logic of the Stoics enables reasoning with propositions of the type "if A then B" – A and B being "atomic propositions," that is, propositions expressing a simple fact, such as "Socrates is mortal" or "the sun is shining." Thus, in the Stoic logic, propositions are atomic propositions connected by conjunctions ("if... then," "or," etc.). This scheme implies a very limited conception of language in which there are only two grammatical categories (atomic propositions and conjunctions) and which does not take into account the fact that an atomic proposition, say "Socrates is mortal," can be decomposed into a subject (Socrates) and a predicate or attribute (mortal).

Unlike that of the Stoics, the logic of Aristotle acknowledges the role of the predicate: the expressions X, Y, and Z that play a role in the reasoning process are precisely predicates (whether "squirrel," "rodent," "mammal," or something else). On the other hand, Aristotle's logic has no place for proper nouns, that is, symbols that refer to individuals or objects such as "Socrates," for, as Aristotle saw it, science does not concern specific cases but only general notions such as "human being," "mortal," and the like. The oft-quoted syllogism "All men are mortal, Socrates is a man, therefore Socrates is mortal" does not belong to Aristotle's logic. To him, the syllogism should read: "All men are mortal, all philosophers are men, therefore all philosophers are mortal." So, in Aristotle's logic, propositions are not formed with a subject and a predicate, but with two predicates and an indefinite pronoun – "all" or "some." It was not until the late Middle Ages that this logic was extended to include individual symbols, such as the proper noun "Socrates," and even thus extended, it remained too coarse to formulate certain mathematical propositions. For example, although Aristotle's logic extended enables us to form the proposition "4 is even" by simply combining the individual

symbol "4" with the predicate "even," it offers no means of forming the proposition "4 is less than 5": whereas, in the first proposition, the predicate "even" applies to a single object, in the second
proposition, the predicate "is less than" applies to two objects – "4"
and "5" – which it puts in relation. For the same reason, we cannot
use this logic to form the proposition "the line L goes through the
point A."

It seems understandable, then, that Greek mathematicians
spurned the logics offered by their contemporary philosophers:
these forms of logic were too meager to express the reasoning processes of the budding disciplines of arithmetic and geometry. For
a very long time, the problem of developing a logic rich enough
for mathematical reasoning provoked very little interest. A few
attempts were made in the seventeenth century – Gottfried Wilhem
Leibniz tried his hand at it – but it was not until 1879 that Gottlob
Frege picked up the problem and put forward a logic of his own. And
these attempts only really began to yield concrete results with Alfred
North Whitehead and Bertrand Russell's theory of types in the early
1900s, and then with David Hilbert's predicate logic in the 1920s.

But let's resume our history of Greek mathematics. Mathematicians may not have had explicit inference rules at their disposal
to conduct their reasoning, but that didn't keep mathematics from
making advances. The inference rules and the grammar of mathematical propositions simply remained implicit until the nineteenth
century. This situation is common in the history of science: when
you are missing a tool, you cobble together makeshift substitutes
that often prefigure the tool itself.

In the case of geometry, however, axioms – that is, facts considered to be established without demonstration and that serve as
premises for proofs – were made explicit by Euclid. The most famous
of these axioms is the axiom of parallels, which, in its modernized
form, can be expressed as follows: through a point not on a given
straight line, one and only one line can be drawn that is parallel to
the given line. We will delve into this axiom in the following.

For a long time, Euclid's treatise, *Elements*, embodied the proto-
typical mathematical method: one states axioms and, using these
along with rules of inference – whether explicit or implicit – one
proves theorems. From the standpoint of this method, reasoning

is the only way to solve mathematical problems. This position is in keeping with the importance the Ancient Greeks, both mathematicians and philosophers, attached to reasoning.

So Greek mathematicians discovered the axiomatic method and, with it, a whole new way of practicing mathematics. They might have tried to understand how this new sort of mathematics followed from Mesopotamian and Egyptian mathematics. If they had, this line of investigation would have led them to look for a way to combine computing and reasoning. But they did no such thing. Quite the contrary – they made a clean sweep of the past and abandoned computing altogether to replace it with reasoning.

For this reason, after the Greeks, computation held hardly a place in the rising edifice of mathematics.

CHAPTER TWO

Two Thousand Years of Computation

ONCE THE AXIOMATIC METHOD had been adopted, reasoning was often spoken of as the one and only tool available for solving mathematical problems. In the discourse they developed about their science, mathematicians hardly ever mentioned computation. This doesn't mean that computing vanished from the practice of mathematics, however. Mathematicians would regularly put forward new algorithms to systematically solve certain types of problems. It seems that the history of mathematics has a bright side – that of conjectures, theorems, and proofs – and a hidden one – that of algorithms.

This chapter will focus on three important points in this history, each set in a different time period, and each raising important issues.

First we will tackle the apparent contradiction between mathematical discourse, which tends to overlook computation, and mathematical practice, which places great weight on it. We will also retrace the transition between the prehistory of mathematics and Ancient Greek mathematics.

Next we will examine the relative parts played in medieval mathematics by the Mesopotamian legacy and by the Greek legacy.

Finally we will explore why so many new geometric figures (the catenary curve, the roulette curve, etc.) appeared in the seventeenth century, whereas ancient geometry focused on only a small number of figures (the triangle, the circle, the parabola, etc.).

EUCLID'S ALGORITHM: REASONING-BASED COMPUTATION

Euclid linked his name not only to geometry and the axiomatic method but also, ironically, to an algorithm that allows the calculation of the greatest common divisor of two integers. It is known as Euclid's algorithm.

The first method for calculating the greatest common divisor of two numbers consists of listing the divisors of each number – successively dividing the number by all smaller numbers and writing down all those for which there is no remainder – and identifying the largest number that appears on both lists. For instance, in order to calculate the greatest common divisor of 90 and 21, we start by listing the divisors of 90 (1, 2, 3, 5, 6, 9, 10, 15, 18, 30, 45, and 90) and those of 21 (1, 3, 7, and 21). Then we observe that 3 is the largest number on both lists. Thus, to verify that 3 is the greatest common divisor of 90 and 21, or even to find out what the greatest common divisor of 90 and 21 is (according to how the problem is phrased), there is no need for reasoning. It suffices to apply this tiresome yet systematic algorithm (which boils down to a simple paraphrase of the definition of greatest common divisor).

Euclid's algorithm enables us to achieve the same result in a less tedious way. It rests on the following idea: in order to calculate the greatest common divisor of two numbers a and b – say, 90 and 21 – we start by dividing the greater number, a, by the smaller, b. If the division works out exactly and produces a quotient q, then $a = b \times q$. In that case, b is a divisor of a, therefore it is a common divisor of a and b, and it is bound to be the greatest one, because no divisor of b can be greater than b itself. As a conclusion, that number is the greatest common divisor of a and b. Now, if the division does not work out exactly but leaves a remainder r, then $a = b \times q + r$. In that case, the common divisors of a and b are also those of b and r. For that reason, we can replace the pair a and b by the pair b and r, which will have the same greatest common divisor. Euclid's algorithm consists in repeating that operation several times until we reach a pair of numbers for which the remainder is zero. The greatest common divisor is the smaller of those two numbers. Thus, when we calculate the greatest common divisor of 90 and 21 using Euclid's algorithm, we first replace the pair (90, 21) by the pair

(21, 6), then by the pair (6, 3), and, finally, 6 being a multiple of 3, the result is 3.

For the numbers 90 and 21, Euclid's algorithm yields a result after three divisions. More generally, whatever numbers we start with, we will reach a result after a finite number of divisions. Because the number a is replaced with the number r, the numbers in the pair whose greatest common divisor we are looking for decrease, and a decreasing series of natural numbers is necessarily finite.

This example shows that, far from turning their backs on computation, the Greeks – among them, Euclid – participated in the devising of new algorithms. It also shows how intricately interwoven reasoning and computing are in mathematical practice. Whereas the first algorithm we discussed required no prior demonstration, in order to elaborate Euclid's algorithm, it was necessary to demonstrate several theorems: first, if the division of a by b works out exactly, then the greatest common divisor of a and b is b; second, if r is the remainder of the division of a by b, then the common divisors of a and b are the same as those of b and r; third, the remainder of a division is always less than the divisor; and last, a decreasing series of natural numbers is necessarily finite. Euclid established those results by reasoning processes similar to those used by the Pythagoreans to prove that a square cannot equal twice another square.

No significant reasoning was needed to build the first algorithm, but this is an exceptional case. More often than not, algorithms are like Euclid's and entail more than merely paraphrasing a definition: in order to elaborate the algorithm, we must conduct a reasoning process.

THALES AND THE PYRAMIDS: THE INVENTION OF MATHEMATICS

The fact that building an algorithm typically requires reasoning causes us to wonder, in retrospect, about Mesopotamian and Egyptian mathematics. How did the Mesopotamians, for instance, conceive a division algorithm without resorting to reasoning? The two peoples must have known an implicit form of reasoning. The fact that, unlike the Greeks, they did not make their reasoning processes explicit – by writing them down on tablets, for example – and that

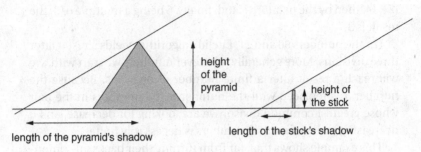

Figure 2.1 (pyramid, stick and shadow)

they probably were not aware of the importance of reasoning in the solving of abstract mathematical problems proves nothing: they may very well have reasoned the way Monsieur Jourdain, Molière's "Bourgeois Gentleman," spoke in prose – "unawares."

The necessity of mathematical reasoning in the building of algorithms has often been remarked upon. More rarely has it been noted that this necessity sheds light on the Greek miracle: the transition from computing to reasoning. Indeed, we can hypothesize that the importance of reasoning dawned on the Greeks precisely as they developed algorithms.

The "first" geometrical reasoning process is generally attributed to Thales. In order to calculate the size of a pyramid that was too high to be measured directly, Thales came up with an idea: he measured the length of the pyramid's shadow, the height of a stick, and the length of the stick's shadow, then proceeded to apply the rule of three (Figure 2.1).

It seems likely that Thales's aim was to devise a new algorithm to calculate the length of a segment; in doing so, he probably realized that he needed to prove that the ratio between the pyramid and its shadow was the same as that between the stick and its shadow. Thus, a theorem was born, the intrinsic value of which was later to be revealed.[1]

---

[1] This result, known today as the "intercept theorem," is referred to in many languages as Thales's theorem. It should not be confused with another result, more commonly known in English as Thales's theorem, which deals with points on a circle.

DISCOURSE AND PRACTICE

Euclid's algorithm raises another question. How can we account for the apparent contradiction between mathematical discourse – which, since antiquity, has set little store by computation – and mathematical practice – where computation plays such an important part? How could the Greeks and their successors claim that reasoning alone mattered while at the same time building algorithms the like of Euclid's?

Let's revisit how Euclid's algorithm helped to find the greatest common divisor of 90 and 21. One way to describe the procedure we carried out would be to say that we replaced the pair (90, 21) first with the pair (21, 6), then with the pair (6, 3), and finally with the number 3, "blindly" applying the algorithm after establishing that it does indeed produce the greatest common divisor of two numbers. From another perspective, we might also justify replacing the pair (90, 21) with the pair (21, 6) by proving that the greatest common divisor of 90 and 21 is the same as that of 21 and 6. All we need to do is to use one of the theorems mentioned earlier, according to which the greatest common divisor of two numbers a and b is the same as that of b and r, where r is the remainder of the division of a by b. Putting things this way makes no mention of Euclid's algorithm: it says simply that we have proved the greatest common divisor of 90 and 21 to be 3, by applying the two theorems stated earlier.

More specifically, Euclid's algorithm not only delivered the result (3), but also enabled us to set out a proof that shows that the greatest common divisor of 90 and 21 is 3. Once the proof is complete, it doesn't matter where it came from – as long as it is there. If the Greeks and their successors thought of computation only as a tool destined to build proofs (and to remain in the shadow of the proofs it helped to produce), then there turns out to be a certain consistency, after all, between their mathematical practice and their mathematical discourse.

POSITIONAL NOTATION

We will now move on to a second important moment in the history of mathematics. It is commonplace to believe that the way we designate mathematical objects, as well as everyday objects, is a detail of little significance. There is no reason to call a lion "lion" or a tiger "tiger." We could have chosen any two other words; the only thing that matters is that we all use the same convention – in other words, "A rose by any other name would smell as sweet." To describe this phenomenon, linguists speak of the arbitrary nature of the sign. Similarly, we could have picked another name than "three" to refer to the number 3 – and another symbol than "3" to write it – it would make little difference, if any. Indeed, other languages do use different words to refer to the same number – "drei," "trois" – and yet this has no impact on mathematics, which remains the same whatever the language.

Taking the thesis of the arbitrary nature of the sign one step further, we might think that it makes no difference whatsoever how we write "thirty-one" – "IIIIIIIIIIIIIIIIIIIIIIIIIIIIIII," "XXXI," "3X 1I," or "31." This is not exactly true.

First, why do we feel the need to develop a special language just for writing down numbers? When naming chemical elements, although we have a specific name for each one – hydrogen, helium – we still use the English language. Similarly, small numbers are each given a specific name – "one," "two," "three" – and even a special symbol – "1," "2," "3" – but, whereas the number of chemical elements is finite, there is an infinity of numbers. One cannot name each of them, for a language can contain only a finite number of symbols and words.

Clearly, assigning a specific symbol to each number is not possible, because there is an infinity of numbers, but another option remains open: designating numbers by combining a finite number of symbols. In other words, instead of specifying a lexicon, one would elaborate a grammar – that is, a language. A grammar is much less arbitrary than a lexicon, and some grammars of the numerical language are far more practical than others when it comes to reasoning and computing. Take a closer look at the different ways of writing "thirty-one": "thirty-one," "IIIIIIIIIIIIIIIIIIIIIIIIIIIIIII," "XXXI," "3X 1I," and "31." The last option, "31," is the best solution, for the very position of the number "3" indicates that it is a tens number; thus, when you write two numbers one above the other, units are automatically lined up with units and tens with tens, which makes simple addition and subtraction algorithms possible. More importantly, with this notation system, the multiplication algorithm is simplified: in order to multiply a number by ten, all you need to do is shift the number to the left and add a 0 at the end.

This positional notation for numbers has its origin in Mesopotamia, where a rough draft of this system was already in use by 2000 B.C. However, the Mesopotamian system was too complicated. The Indians were the first to simplify it. Then, in the ninth century, the Indian version of positional notation spread to the Arab world thanks to a book written by Muhammad ebne Mūsā al-Khwārizmī (from whose name the word "algorithm" is derived) called *Al-Jabr wa-al-Muqabilah* ("Book on Integration and Equation"). The system then reached Europe in the twelfth century. Mathematicians of the Middle Ages thus benefited from a double legacy: they inherited a lot from the Greeks, but also from the Mesopotamians, who handed down to them the all-important positional notation system. These mathematicians then spent many centuries developing and perfecting algorithms.

The discovery of the axiomatic method did not oust computation. On the contrary, computation thrived, through the Mesopotamian legacy, to become a key preoccupation in the eyes of medieval mathematicians.

CALCULUS

Having dealt with Euclid's algorithm and with algorithms designed to carry out arithmetic operations, we now move on to a third crucial event in the history of mathematics: the development of calculus. This branch of mathematics appeared in the seventeenth century with the works of Bonavantura Cavalieri, Isaac Newton, Gottfried Wilhelm Leibniz, and others. Its roots, however, go back much further: during antiquity, two discoveries of Archimedes's laid the groundwork for the invention of calculus. One of these discoveries concerns the area of the circle and the other, the area of the parabolic segment.

It is a well-known fact, today, that the area of a circle is obtained by multiplying the square of its radius by 3.1415926.... Archimedes did not get quite this far, but he did prove that, in order to

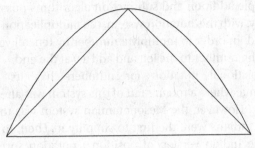

Figure 2.2a

calculate the area of a circle, you need to multiply a number bounded between $3 + 10/71 = 3.140...$ and $3 + 10/70 = 3.142...$ by the square of the circle's radius – in other words, he discovered the first two decimals of the number π. His work on the area of the parabolic segment was even more successful, as he reached an exact result: he correctly established that the area of a parabolic segment equals four-thirds the area of the triangle inscribed within that segment (Figure 2.2a).

In order to achieve that feat, Archimedes decomposed the parabolic segment into an infinity of successively smaller triangles, the areas of which he added up (Figure 2.2b).

If you take the area of the triangle inscribed within the parabolic segment as a unit, the area of first triangle is, by definition, 1. It can be proved that the two triangles on its sides have a total area of $1/4$, then that the area of the next four triangles is $1/16$, and so on. The total area of each set of triangles equals one fourth that of the

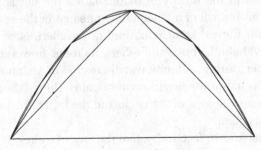

Figure 2.2b

previous set. The area of the parabolic segment is obtained by adding up the areas of all these triangles: $1 + (1/4) + (1/4)^2 + (1/4)^3 + \cdots$, and the sum of this infinity of numbers yields a finite result: 4/3. Archimedes, who was understandably reluctant to add an infinity of numbers, contented himself with examining the finite sums $1, 1 + (1/4), 1 + (1/4) + (1/4)^2, \ldots$, which represent the areas of the polygons inscribed within the parabolic segment, and which are therefore all smaller than the area of the parabolic segment. He showed that the area of the parabolic segment could not be smaller than 4/3, for it would then be smaller than that of one of the inscibed polygons, which is impossible. Archimedes then used another argument (resting, this time, on the circumscribed polygons) to demonstrate that the area of the parabolic segment could not be greater than 4/3. If the area of the segment could be neither smaller nor greater than 4/3, it was bound to equal 4/3.

This reasoning detour – through finite sums and bounds above and below – was bypassed in the sixteenth century when mathematicians, among them Simon Stevin and François Viète, started adding up infinite lists of numbers. However, even this simplified variant of Archimedes's reasoning process still required proving the total area of each set to be a fourth of the previous set's area – which was quite a tour de force! So, until the seventeenth century, establishing the area of geometric figures remained a headache.

In the seventeenth century, after René Descartes introduced the notion of coordinate, it became possible to describe many curves by an equation. For instance, the parabola in Figure 2.2 would be described as $y = 1 - x^2$ (Figure 2.3).

Knowing this equation, we might look for a way of calculating the area of the parabolic segment – that is, the area bounded between the curve and the horizontal axis – without decomposing the segment into triangles. In fact, one of the greatest discoveries of seventeenth century mathematics concerned the method used to calculate the area of a figure delimited by a curve of known (and relatively simple) equation.

The first step toward this discovery consisted in establishing a link between the notion of curve-enclosed area and that of derivative.

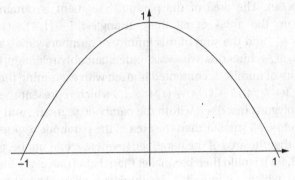

Figure 2.3

Take a function – say, the one that assigns to the number x the value $x - x^3/3$. The value of this function at $x + h$ is $(x + h) - (x + h)^3/3$. Simple algebraic reasoning shows that the difference between the value of this function at $x + h$ and its value at x is $h - x^2h - xh^2 - h^3/3$. The function's "difference quotient" between x and $x + h$ can be obtained by dividing this quantity by h, which produces the following result: $1 - x^2 - xh - h^2/3$.

A function's instantaneous growth rate at a point x (in other words, its "derivative" at x) can be obtained by watching what happens to this difference quotient when h tends to 0: in our example, the last two terms disappear, which leaves us with $1 - x^2$.

However, to find the derivative of the function that assigns to x the value $x - x^3/3$, this reasoning process is not necessary. Instead, because it can be proved that the derivative of the sum of two functions equals the sum of their derivatives; all we need to do is to determine the derivative of x and that of $-x^3/3$ and add them. And we can demonstrate that multiplying a function by a fixed quantity, or constant, amounts to multiplying its derivative by the same constant. Thus, in order to determine the derivative of $-x^3/3$, we determine the derivative of x^3 and multiply it by $-1/3$. Finally, to find the derivative of x and of x^3, we need only remember that the derivative of x^n is $n \times x^{n-1}$. Therefore, the derivative of $x - x^3/3$ equals $1 - x^2$.

What is the difference between these two ways of calculating the derivative of $x - x^3/3$? In the first method, a reasoning process was necessary – it was simple enough, but it did require some thinking. In the second method, we obtained the derivative of $x - x^3/3$ systematically, by applying the following rules:

- The derivative of a sum is the sum of the derivatives;
- Multiplying a function by a constant multiplies its derivative by the same constant;
- The derivative of x^n is nx^{n-1}.

Once these three rules have been proved correct, simple computation is all we need to obtain the derivative of many functions. The algorithm that allows us to calculate a function's derivative does not apply to numbers, but rather to functional expressions. What is more, it does not apply to all functional expressions, but only to those that can be obtained from x and constants by means of addition and multiplication – in other words, it applies only to polynomials. Other algorithms, more general yet similar in nature, are applicable to richer languages that include, among other things, exponential and logarithmic functions as well as trigonometric functions.

The function that assigns to x the value $1 - x^2$ is the derivative of the function that assigns to x the value $x - x^3/3$. The function that assigns to x the value $x - x^3/3$ is called a "primitive," or "antiderivative," of the one that assigns to x the value $1 - x^2$. It can be proved that the latter has several primitives, all of which are obtained by adding a constant to the primitive above.

From the rules that enable us to calculate a function's derivative, we can easily build an algorithm to compute primitives:

- Taking the sum of primitives of functions gives a primitive of their sum;
- A primitive of a constant multiple of a given function can be obtained by multiplying one of its primitives by the same factor;
- A primitive of x^n is $x^{n+1}/(n + 1)$.

By systematically applying these rules, we can calculate a primitive of $1 - x^2$, namely $x - x^3/3$.

Now consider once more the problem of areas. The fundamental theorem of calculus establishes a connection between the notion of area and that of primitive. Indeed, if $F(x)$ is the function that assigns

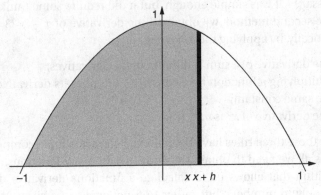

Figure 2.4

to x the area of the part of the parabolic segment situated left of the vertical line at x, it can easily be shown that the derivative of the function F is the function that assigns to x the value $1 - x^2$ (Figure 2.4).

In other words, the function F is a primitive of $1 - x^2$: to x, it assigns $x - x^3/3$ plus a fixed value. Because the function F equals 0 at $x = -1$, this fixed value is bound to be $2/3$ and F is the function that, to x, assigns $x - x^3/3 + 2/3$. The area of the parabolic segment is the value of this function at 1, namely $4/3$. Thus we reach the same result as Archimedes without decomposing the parabolic segment into triangles. So, in order to calculate the area of the parabolic segment bounded by the curve $y = 1 - x^2$, there is no need to develop a complex reasoning process, no need to decompose this segment into triangles, and no need to calculate the area of those triangles. It suffices to calculate a primitive of $1 - x^2$ by applying an algorithm, to adjust the constant so that this primitive equals 0 at -1, and to take its value at 1.

Like the algorithm used to calculate derivatives, the algorithm for calculating primitives applies only to polynomials. As before, there exist more general algorithms that apply to richer languages; however, this specific algorithm cannot be generalized as easily as the one used to calculate derivatives. As a result, over more than three centuries, the calculation of primitives was little more than a combination of algorithms and sleights of hand that required some skill, now in the field of computation, now in that of reasoning. Only in the twentieth century did the theory of algorithmic integration become systematized, thanks to the development of computer algebra systems (of which more later).

In the seventeenth century, as the notions of derivative and primitive developed along with their corresponding algorithms, determining many areas, volumes, lengths, centers of gravity, and so forth became simple computing tasks. When solving this particular type of problem was made systematic, and hence easier, suddenly vast new mathematical territories lay open to exploration. The mathematicians of antiquity had computed the areas of a few figures; those of the seventeenth century went much further. Calculating the area delimited, between -1 and 1, by a complex curve (such as the curve defined by the equation $y = 2 - x^2 - x^8$) would have been extremely difficult for the Greeks, but it was child's play for seventeenth century mathematicians: all they had to do was to calculate the primitive of $2 - x^2 - x^8$ that equals 0 at -1 (so $2x - x^3/3 - x^9/9 + 14/9$) and to take its value at 1, namely $28/9$. Such algorithmic tools are a real asset; without them, tackling such problems would be difficult. In addition to these new tools, new geometric figures, which would have been too hard to study without these tools, appeared in the seventeenth century.

This incursion of algorithmic methods into geometry – the holy of holies of the axiomatic method – left a deep imprint on this branch of mathematics. Its very name bears the marks: it is never referred to as "integral theory," but always as "calculus."

PART TWO

The Age of Reason

CHAPTER THREE

Predicate Logic

IN THE PREVIOUS CHAPTERS, we focused on a few key events illustrating the enduring problem of computation throughout the history of mathematical practice. We could have explored this issue at greater length: we could have mentioned Pascal's triangle, Gauss–Jordan elimination, and so forth. We choose, however, to leave this arduous task to historians and proceed to the heart of this book: the metamorphoses of computation in the twentieth century.

In the twentieth century, the notion of computation developed in tandem with that of reasoning, so we will approach the metamorphoses of computation by way of a brief detour through the history of reasoning. So far, we have traced this history up to the point at which the Stoics and Aristotle developed their separate logics – neither of which turned out to be rich enough to express mathematical propositions. In the centuries that followed, little progress was made on conceiving a grammar fit for mathematical reasoning – despite some valiant attempts, for example, by Leibniz – until the late nineteenth century, when Gottlob Frege turned his hand to it. Ironically, Frege's motives in taking on this challenge were philosophical rather than mathematical: his aim was to shed light on a specific point of Emmanuel Kant's philosophy the better to refute it.

SYNTHETIC A PRIORI JUDGMENTS

The propositions "triangles have three angles" and "the earth has a satellite" are both true, but for different reasons. The fact that triangles have three angles is inherent in the definition of the word "triangle," whereas nothing in the definition of the word "earth"

31

implies that the earth has a satellite. In other words, one cannot imagine a triangle having four or five angles, but it is quite possible to imagine the earth having no satellite – after all, Mercury and Venus have none. Kant calls "analytic" the judgment according to which a proposition is necessarily – that is, by definition – true; he calls "synthetic" the judgment according to which a proposition is true, but not by definition. Thus, in declaring the proposition "triangles have three angles" to be true, one makes an analytic judgment; in declaring the proposition "the earth has a satellite" to be true, one makes a synthetic judgment.

Kant also distinguishes between a priori and a posteriori judgments. A judgment is said to be a priori when it takes place solely in someone's mind; it is called a posteriori when it requires some interaction with nature. For instance, asserting that triangles have three angles requires nothing more than thinking – the judgment is formed in the mind. However, even a great deal of thinking does not allow someone to state with certainty that the earth has a satellite – in order to do so, at some point he or she needs to observe the sky.

These examples may mislead us into thinking that these two oppositions coincide or, in other words, that analytic judgments are always a priori and synthetic judgments always a posteriori. This is not the case. Some judgments are both synthetic and a priori, for it is not always necessary to carry out experiments in order to know nature. The most famous example of a synthetic a priori judgment is the proposition "I am." Nature has existed without me for a long time, it will survive me for a long time, too, and it would probably have carried on existing without my ever coming into being. It cannot be said that I exist by definition. Therefore, the judgment according to which "I am" is synthetic. However, I do not need to observe nature in order to convince myself of my own existence. If I were trying to ascertain that kangaroos exist, I would have to observe Australian wildlife, but, as far as my own existence is concerned, I think, therefore I am. The fact that time exists is also a synthetic a priori judgment. Time does not exist by definition, and there is no necessity for things to ever move or change. Still, we don't need to look around us to know that time exists: our consciousness evolves in time and this suffices to make us aware of the existence of time.

To this point, Frege agrees with Kant. What he disagrees with is this: Kant claims that all mathematical judgments belong in the synthetic a priori category. It seems obvious enough that mathematical judgments are a priori: to construct a proof, as the Pythagoreans did to demonstrate that a square cannot be twice another square, or that the sum of the angles in a triangle is always 180°, you do not require a microscope or a telescope – indeed, you have no need to look around yourself at all. On the other hand, Kant believes that the judgment according to which the proposition "triangles have three angles" is true differs in kind from the judgment according to which the proposition "the sum of the angles in a triangle is 180°" is true: having three angles is part of the definition of the notion of triangle; the fact that the sum of these angles is 180° is not. So the first judgment is analytic, but not the second – it is bound to be synthetic. Therefore, the second judgment is synthetic a priori. Similarly, for Kant, the fact that $2 + 2 = 4$ does not derive from the definition of either 2 or 4 – it is yet another synthetic a priori judgment.

FROM THE NOTION OF NUMBER TO THOSE OF CONCEPT AND PROPOSITION

Frege aimed to prove that, contrary to what Kant claimed, the fact that $2 + 2 = 4$ is an implicit part of the definition of the integers, for even if this fact is not explicitly stated in the definition, it can be inferred from it by reasoning. Frege therefore suggested that the role of reasoning was to make explicit the properties implicitly contained in the definitions of mathematical concepts, even the people who provided these definitions were unaware of the implicit properties they held. A person's own definitions – just like his or her actions – can entail unforeseen yet inevitable consequences.

In order to defend his thesis, Frege needed to give a definition of the integers, to make the rules of inference explicit, and to show how, with the aid of reasoning, the truth of a proposition such as $2 + 2 = 4$ could be obtained from the definition.

The importance that Descartes and later Kant, following in the tradition of Plato and others, attached to the mind's faculty for establishing facts about nature without resorting to anything outside itself naturally invites us to wonder about the mechanisms that allow the mind to do so and, more specifically, about the rules of reasoning – something that, surprisingly enough, Plato, Descartes, and Kant all overlooked to some extent.

Frege, however, tackled this issue. In this respect, although his motives were philosophical, he can be compared to Richard Dedekind and Giuseppe Peano, two mathematicians of the late nineteenth century who also strove to define the natural numbers. Indeed, whereas the first definitions of the notions of "point" and "straight line" date back to Euclid, as do the first geometric axioms, the natural numbers were not defined nor were arithmetic axioms set out until the end of the nineteenth century. Why not? Perhaps because when we already know rules of computation that allow us to establish that $2 + 2 = 4$, developing axioms to prove it doesn't seem especially urgent. Since the days of the Pythagoreans, we have known that reasoning is necessary to arithmetic; however, until the late nineteenth century, the role of reasoning may have appeared less central to arithmetic than to geometry.

To construct his definition of the integers, Frege drew inspiration from an idea David Hume had sketched more than a century earlier. Frege suggested that we define integers as sets of sets. The number 3, for example, can be defined as the set { {Athos, Porthos, Aramis}, {Olga, Masha, Irina}, {Huey, Dewey, Louie}, ... } containing all sets of three elements. So, in order to define the notion of integer, it is necessary first to define the notion of set.

For Frege, the notion of "set" was intertwined with that of "concept": the set containing all roses is nothing more than the concept "to be a rose," and a concept is defined by a proposition. Thus did Frege reopen an issue that had been more or less dormant since antiquity, namely, clarification of the grammar of mathematical propositions and rules of inference.

FREGE'S LOGIC

Frege borrowed a few concepts from the Stoics – for example, in Frege's logic, a proposition is formed of atomic propositions interconnected by coordinating conjunctions ("and," "or," "no," "if...then," etc.). Unlike the Stoics and like the logicians of the Middle Ages, Frege decomposed atomic propositions; however, he decomposed them not into two elements (a predicate and its subject), but into a "relational predicate" and the several complements it links (among which is the subject). Take the atomic proposition "4 is smaller than 5": for the Stoics, this proposition was indecomposable; for the logicians of the Middle Ages, it was decomposable into a subject, "4," and its predicate, "is smaller than 5"; for Frege, it is made up of a relational predicate, "is smaller than," and the two complements it connects, "4" and "5."

Frege's logic (1879), like Aristotle's, can express the fact that a predicate applies not just to one particular object but to all possible objects, or to a few objects, without specifying which ones. In traditional grammar, as in Aristotle's logic, we replace the subject or a complement with an indefinite pronoun – "all" or "some." Thus, the proposition "all are mortal" is formed in the same way as the proposition "Socrates is mortal" – we just replace the noun "Socrates" with the pronoun "all." This mechanism is a source of ambiguity in natural languages, for with a relational predicate, the proposition "everybody loves somebody" can mean either that there is one person who is loved by all others, or that everybody loves a person without that person necessarily being the same for all. In natural languages, we work around that imperfection by adding specifics when the context alone does not clear up the ambiguity. In order to clarify inference rules, however, we must give different forms to proposals that have different meanings.

For this purpose, Gottlob Frege and Charles Sanders Peirce deployed an invention of the sixteenth century algebraists, among them François Viète: the notion of variable. First, instead of applying the relational predicate "is smaller than" to nouns "4" and "5" or to indefinite pronouns, we apply it to variables "x" and "y," producing the proposition "x is smaller than y." Second, we specify whether those variables are universal or existential, using the propositions "for all x" or "there exists x". Thus, the Aristotelian proposition "all men are mortal" can be decomposed, in Fregian logic, into "for all x, if x is a man, then x is mortal." "Everybody loves somebody" can be translated as either "for all x, there exists y such that x loves y" or "there exists y such that for all x, x loves y."

The ambiguity of a sentence such as "everybody loves somebody" or "every number is smaller than a certain number" is cleared up by the order of the quantifiers in the proposition. Thus, the proposition "for all x, there exists y such that x is smaller than y" means that, for every number, there exists a bigger number, which is true, whereas the proposition "there exists y such that for all x, x is smaller than y" means that there exists a number bigger than all the others, which is false.

Once the grammar has been enriched with relational predicate symbols, variables, and the quantifiers "for all" and "there exists," inference rules are not hard to express. One of them, with which the Stoics were familiar, allows us to deduce the proposition B from the propositions "if A then B" and A. Another allows us to infer the proposition A from the proposition "A and B." The most interesting inference rules, however, are those that involve quantifiers: for example, there is a rule that allows us to deduce, from the proposition "for all x, A," the proposition A in which the variable x is replaced with some term. Using this rule, from the proposition "for all x, if x is a man, then x is mortal," we can infer "if Socrates is a man, then Socrates is mortal." Then, knowing that "Socrates is a man," we can deduce, using the first rule, that "Socrates is mortal."

So, in Frege's logic, we first define the notion of set, then that of integer, then the numbers 2 and 4, then the addition operation, and, finally, we can prove the proposition "$2 + 2 = 4$." This shows that the truth of the last proposition is a consequence of the definition of the integers and of the addition operation and, therefore, that the judgment according to which this proposition is true is analytic, and not synthetic as Kant thought.

THE UNIVERSALITY OF MATHEMATICS

In devising his logic, Frege achieved a double feat. He not only succeeded in synthesizing the logic of Aristotle with that of the Stoics, developing a logic richer than those of the Greek philosophers. By deriving the notion of number from that of set, he also gave a definition of the natural numbers and proved the proposition "$2 + 2 = 4$" to be true.

What is true for the natural numbers is also true for many other mathematical concepts: using Frege's logic, we can define just about all the notions we want and prove nearly all the theorems we know.

So Frege's logic has two sides. On one hand, it simply expresses the rules of reasoning and states axioms about general notions, such as sets or concepts. In this respect, it is in the tradition of the logics of Aristotle and the Stoics, and we can label a proof "logical" if it can be formulated in Frege's logic. On the other hand, Frege's logic makes it possible to formulate all mathematical reasoning processes. It thus continues the tradition of Euclidian geometry, and we can label a reasoning process "mathematical" if it can be formulated in Frege's logic.

Frege's logic reveals that the adjectives "logical" and "mathematical" are synonyms. In other words, there is no specifically *mathematical* reasoning: any logical reasoning can be described as mathematical. Therefore, it is no longer appropriate to define mathematics in terms of its objects – numbers, geometric figures, and so forth. Rather, we should define the subject in terms of the way it describes and deals with these objects or, in other words, in terms of the logical proofs it constructs. The specificity of mathematics, which was assumed to lie in its objects, suddenly fades, and mathematics emerges universal.

At the beginning of the twentieth century, Bertrand Russell underlined the importance of Frege's discovery. The universality of mathematics is essential, and those who deny mathematics any relevance in the social sciences should bear it in mind. Indeed, asserting that mathematics is not an appropriate tool for studying human behavior amounts to claiming that logic and reasoning are not adequate tools for this purpose, which amounts to denying the very existence of the social sciences. Of course, this position can be defended, but anyone holding it must keep in mind that, since Frege and Russell, it goes far beyond the commonplace idea that numbers and geometric figures are not suitable tools for studying human behavior.

PREDICATE LOGIC AND SET THEORY

Frege's logic was still imperfect and these successes were followed by a few setbacks. For instance, in this logic, it is possible to prove both a thing and its opposite. And in mathematics, as in everyday life, such a contradiction is the sign of a mistake.

The first paradox in Frege's logic was found by Cesare Burali Forti in 1897; it was then simplified and popularized by Bertrand Russell in 1902. In Frege's logic, some sets are elements of themselves – the set of all sets, for instance. With this observation in mind, we can form a set R of all sets that are *not* elements of themselves. We can then prove that the set R is an element of itself, and also that it is not. Frege's logic was clearly a first draft that needed to be improved.

And improved it was. Russell in 1903 was the first to offer a revised version of Frege's logic, namely type theory, which he developed in the years that followed with Alfred North Whitehead. Russell and Whitehead's type theory creates a hierarchy of types, then assigns each mathematical entity to a type: entities that are not sets (atoms) are designated as type 0, sets of entities of type 0 (sets of atoms) are called type 1, sets of entities of type 1 (sets of sets of atoms) are called type 2, and so forth. Objects of a given type are built exclusively from objects of preceding types so that the proposition "x is an element of y" can only be formed when the type of y equals that of x plus one. Consequently, there is no set of all sets, because it is forbidden to put together sets of atoms and sets of sets of atoms, nor can we speak of sets that are not elements of themselves. Russell's paradox is eliminated.

Another shortcoming of Frege's logic – shared by Russell and Whitehead's – is that it mixes notions of logic with notions of set theory. An inference rule – say, the rule that allows us to deduce the proposition B from the propositions "if A then B" and A – can be useful in different fields of inquiry. By contrast, the axiom stating that "only one straight line can pass through two points" is valuable only in geometry, for it is expressed explicitly in terms of points and straight lines. Similarly, the axiom "if A and B are two sets, there exists a set whose elements are the elements common to A and B" is useful only in set theory.

Traditionally, inference rules, which are valid no matter what object they are applied to, are contrasted with axioms, which are valid only within the framework of a particular theory. The fact that rules of inference are object neutral has a scientific name: the "ontological neutrality of logic." Frege's logic, like Russell's later version, is flawed because it rests on inference rules tied to the specific notion of set (or concept), and whereas this notion may have seemed general in Frege's days, Russell's paradox proved it to be just another notion requiring its own axioms. So, in the 1920s, Russell's type theory was in turn simplified: David Hilbert stripped it of everything specific to the notion of set and created "predicate logic," which remains the framework of logic to this day. Since then axioms specific to the notion of set, as they were formulated by Ernst Zermelo in 1908, constituted just one theory among many others, namely set theory.

The separation between predicate logic and set theory undermines Russell's thesis of the universality – or lack of specificity – of mathematics. It is predicate logic that appears to be universal; within predicate logic, if you wish to practice mathematics, it is necessary to call upon axioms taken from set theory. It is therefore possible to conceive of a logical reasoning process that obeys the rules of predicate logic and yet rests on axioms other than those of set theory; the same cannot be said of mathematical reasoning.

Actually, these assertions call for some qualification. A theorem proved by Kurt Gödel in 1930 (but not Gödel's famous theorem) shows that any theory can be translated into set theory. Euclidian geometry, which a priori rests on a different set of axioms than set theory, can nevertheless be translated into set theory. This theorem revives Russell's thesis by conferring universality and ontological neutrality on set theory itself.

THE PROBLEM OF AXIOMS

Besides endangering Russell's thesis, the separation of predicate logic and set theory has another major drawback: it imperils Frege's whole philosophical project to define the notion of a natural number from purely logical notions, then to show that the proposition "$2 + 2 = 4$" follows from that definition. In predicate logic, without axioms it is impossible to define the integers so as to make this proposition provable. As soon as we introduce axioms – such as those of set theory – it becomes possible. Around the same time that Frege put forward his axioms, Peano devised some of his own, namely axioms of arithmetic. These also made it possible to prove the proposition "$2 + 2 = 4$" and, more generally, to demonstrate all known theorems concerning integers, only in a simpler way.

The question of the analyticity of mathematical judgments thus shifts the focus onto the notion of axiom. The judgment according to which a proposition is true can be described as analytic if it rests on a proof. But what if this proof relies on axioms? And what about judgments that axioms themselves are true: are those analytic or synthetic judgments? Behind this jargon, an old question about axioms rears its head: why do we accept the assertion of axioms? Meticulous demonstration is the creed of mathematics, so why do we accept, without any justification, the axiom "there is only one straight line through two points"?

At the beginning of the twentieth century, Henri Poincaré offered an answer to this question. A story will help to explain it. While sailing towards the New Hebrides, an explorer discovers an island completely cut off from the world, yet inhabited by English-speaking men and women. At first, the explorer wonders at this fact, but he soon comes across an explanation: according to a local myth, the natives are descended from sailors whose ship, sailing under the British flag, was wrecked not far from the island, a very long time ago. Yet the explorer is in for a bigger surprise: a few days later, he learns that the natives feed on flying fish! Indeed, they claim to hunt a fish that has two wings, two legs and a beak, that builds nests and sings pretty tunes. The explorer tries to explain to the natives that fish live in the sea, are covered in scales and have gills, and that they most certainly do not sing, which causes great mirth among his listeners. Soon, the reason for the misunderstanding dawns on the explorer: the language spoken on the island has evolved a lot since the shipwreck, as has the language spoken in England, so that when the explorer arrives, centuries later, the word "fish" has come to mean, in the natives' language, what the word "bird" means in standard English. This isn't unusual: the explorer knows, for instance, that differences in vocabulary exist between Britain and the United States.

How can the explorer understand the meaning of the word "fish" in the natives' language? He can ask them to show him a fish. If they show a bird, a snake, or a frog, he will know what this word refers to in their language. Unfortunately, this method only works when dealing with concrete things. It is no help with words such as "solidarity" or "commutativity." A second approach is to ask the natives to define the word "fish" or to look up it in one of their dictionaries – but this only works if he already knows what the words used in the definition mean; otherwise, he will have to look up their definitions, and so on, endlessly. A third approach is to ask the natives what propositions concerning the word "fish" they hold to be true. This is the method the explorer used when he grasped the natives' meaning of the word "fish" once he learned that they held true the propositions "fish have two wings," "fish fly in the sky," and so forth.

This is how philosophers define the word "meaning": the meaning of a word is the body of true propositions that contain the word. As a consequence, and contrary to what dictionaries suggest, it is impossible to define the meaning of an isolated word. The meaning of every word is defined simultaneously by the body of all true propositions in the whole language. The truth of the proposition "birds fly in the sky" contributes to defining the word "bird," and also the words "sky" and "fly." To be more precise, it is not actually the body of true propositions that defines the words of a language, for that body is infinite and quite complex; it is the criteria by which we establish that a proposition is true – in the case of mathematical language, those criteria would be axioms and rules of inference.

We can now return to the question about axioms raised at the start of this discussion and give it an answer. We accept the axiom "there is only one straight line through two points" without question because it is part of the definitions of the words "point," "straight line," "through."

This is the explanation that Poincaré proposed, restricted to the language of geometry, and it is far more satisfying than the traditional answer that the truth of this proposition becomes evident as soon as we know the meanings of the words "point," "straight line," "through," and so forth. Thanks to Poincaré, we now understand that this axiom is not a proposition that miraculously seems obvious once we know the meanings of the constituent words; rather, we accept this axiom because it is part of the definitions of those words.

This concept of definition addresses an old problem posed by the definition of the notion of point in Euclid's *Elements*. Euclid gives a rather obscure definition: according to him, a point is that which has no parts. Then he moves on to introducing axioms and proving theorems without ever using this definition. It is natural to wonder why he bothered giving it in the first place. To Poincaré's mind, this definition has no use. The real definition of the notion of point does not lie in this obscure sentence but in the axioms of geometry.

THE RESULTS OF FREGE'S PROJECT

With the notion of axiom clarified, we can attempt to take stock of Frege's efforts. Certainly, contrary to what Frege hoped, it is not possible to define the natural numbers and arithmetic operations and prove the proposition "2 + 2 = 4" using predicate logic without resorting to axioms. However, if we agree to extend the notion of definition and regard these axioms as implicit definitions of the concepts involved, then Frege's project was crowned with success. The proposition "2 + 2 = 4" is a consequence of the axioms of arithmetic, and it is therefore a consequence of the definitions of the natural numbers and their arithmetic operations that these axioms express.

At the conclusion of this mathematical journey from Frege to Hilbert we can see that the concept of analytic judgment was still hazy in Kant's days. There are, in fact, different degrees of analyticity. The judgment that a proposition is true can be called "analytic" when that proposition is provable in predicate logic without the aid of axioms – in which case mathematical judgments are not analytic. Such a judgment can also be called "analytic" when the proposition is provable in predicate logic with the aid of the axioms that constitute the implicit definitions of the concepts used – in which case all mathematical judgments are analytic. As we shall see, other notions of analyticity will be added to these.

Another consequence of this journey was the building of predicate logic. A synthesis of Stoic and Aristotelian logics was achieved, finally producing a logic rich enough to express any mathematical reasoning process – as long as it included a few axioms. The construction of this logic was an important advance toward understanding the nature of reasoning – probably the most important since antiquity. This clarification of the nature of reasoning has had major implications for the relationship between reasoning and computation.

The Genesis of Predicate Logic: 1879–1928

1879: Frege's Logic
1897: Burali Forti's paradox
1902: Russell's paradox
1903: Russell's type theory (further developed with Whitehead)
1908: Zermelo's axioms of set theory
1928: Hilbert's predicate logic reaches its final form

CHAPTER FOUR

From the Decision Problem to Church's Theorem

AT THE BEGINNING OF THE TWENTIETH CENTURY, two theories about computation developed almost simultaneously: the theory of computability and the theory of constructivity. In a perfect world, the schools of thought that produced these theories would have recognized how their work was connected and cooperated in a climate of mutual respect. Sadly, this was not the case, and the theories emerged in confusion and incomprehension. Not until the middle of the century would the links between computability and constructivity finally be understood.

To this day, there are traces of the old strife in the sometimes excessively ideological way in which these theories are expounded, both of which deserve a dispassionate presentation. After our call for unity, we must nevertheless concede that, although both schools of thought developed concepts that turned out to be rather similar, the problems they set out to solve were quite different. Therefore, we will tackle these theories separately. Let's start with the notion of computability.

THE EMERGENCE OF NEW ALGORITHMS

As they each in turn attempted to clarify inference rules, Frege, Russell, and Hilbert contributed to the elaboration of predicate logic. Predicate logic, in keeping with the axiomatic conception of mathematics, consists of inference rules that enable proofs to be built, step by step, from axiom to theorem, without providing any scope for computation. Ignoring Euclid's algorithm, medieval arithmetic algorithms, and calculus, predicate logic signaled the return to the axiomatic vision passed down from the Greeks that turns its back on computation.

In predicate logic, as in the axiomatic conception of mathematics, a problem is formulated as a proposition, and solving the problem amounts to proving that the proposition is true (or that it is false). What is new with predicate logic is that these propositions are no longer expressed in a natural language – say, English – but in a codified language made up of relational predicate symbols, coordinating conjunctions, variables, and quantifiers. It then becomes possible to classify problems according to the form of the propositions that express them. For example, problems of determining the greatest common divisor of two numbers are expressed in propositions of the form "the greatest common divisor of the numbers x and y is the number z."

Euclid's algorithm changes status with predicate logic and the fact that problems of determining the greatest common divisor are characterized by the form of the propositions that express them. It can be seen as an algorithm enabling us to decide whether a proposition of the form "the greatest common divisor of the numbers x and y is the number z" is provable or not. This change of status is slight – an algorithm for calculating the greatest common divisor of two numbers becomes an algorithm for deciding whether a proposition of this form is provable or not – but it is significant.

In the early twentieth century, many algorithms were developed to determine whether certain sets of propositions of predicate logic were provable or not. Some of these algorithms were already known – Euclid's algorithm (which concerns the provability of all propositions of the form "the greatest common divisor of the numbers x and y is the number z") and the addition algorithm (which concerns the provability of all propositions of the form "$x + y = z$") are two examples. Others soon swelled their ranks: in 1929, Mojzesz Presburger conceived an algorithm to assess all "linear arithmetic" propositions, that is, all propositions in the fragment of number theory that contains addition but omits multiplication. This algorithm indicates, for instance, that the proposition "there exist x and y such that $x + x + x = y$" is demonstrable whereas the proposition "there exist x and y such that $x + x = y + y + 1$" is not. Conversely, in 1930, Thoralf Skolem devised an algorithm to determine the provability of all propositions in number theory involving multiplication but not addition. The same year, Alfred Tarski put forward a similar algorithm applicable to all propositions in the theory of real numbers with both addition and multiplication.

These algorithms were more ambitious than Euclid's and than the addition algorithm, because the sets of propositions to which they apply are much larger – indeed, vast. The addition algorithm is used to decide the provability of propositions of the form "$x + y = z$." Presburger's algorithm allows us to determine the provability of propositions not just of this form, but of the forms "there exist x and y such that $x + x + x = y + y + 1$," "for all x, there exists y such that $x + x + x + x = y + y$," "there exist x and y such that $x + x = y + y + 1$," and so forth.

Tarski's algorithm is more ambitious still, for all problems in Euclidian geometry can be boiled down to problems that concern real numbers formulated with addition and multiplication. A consequence of Tarski's theorem, then, is that all geometry problems can be solved by computation. Whereas the Greeks had introduced reasoning to solve problems – especially in geometry – that they could not solve by computing, Tarski showed that, in the case of geometry at least, moving from computation to reasoning is after all unnecessary, because an algorithm that the Greeks had not foreseen can be substituted for reasoning.

THE DECISION PROBLEM

A question naturally arises: can algorithms replace reasoning in mathematics as a whole? The Greeks had introduced reasoning to deal with problems they could not solve by computation alone. With hindsight, however, mathematicians of the early twentieth century wondered whether there might exist an algorithm, unknown to the Greeks, that can be substituted for reasoning in all of mathematics, as was the case in geometry.

Is there such a thing as an algorithm that can be applied to any proposition in order to determine whether that proposition is provable in predicate logic? In the twenties, Hilbert called this question the "decision problem."

When a problem can be solved by an algorithm, it is called "decidable" or "computable." A function f – say the function that, to a pair of numbers, assigns the greatest common divisor of those numbers – is said to be computable when there is an algorithm that, using the value x, calculates the value $f(x)$. Hilbert's decision problem can therefore be rephrased as follows: is it possible to compute the function that, to a proposition, assigns the value 1 or 0 according to whether that proposition is provable or not?

With this problem, predicate logic becomes more than a mere set of rules to be applied whenever we construct a mathematical proof, it becomes an object of study in itself. This shift creates a radical distinction between the way mathematicians thought in Hilbert's day and the way they thought in the time of Euclid. For the Greeks, the objects of their study were numbers and geometric figures, and reasoning was a method. For mathematicians of the twentieth century, reasoning itself was an object of study. The emergence of predicate logic, which elucidates the rules of inference, was a essential step in this process. The Greeks were content with implicit inference rules, but the mathematicians of the twentieth century demanded that these rules be explicitly defined.

Hilbert's motive in suggesting this return to computation was not solely practical. Frege had put forward a contradictory logic in which one could prove a thing and its opposite, which led Russell, and later Hilbert, to make alterations to it. The result they reached, namely predicate logic, did not seem contradictory for, at the time, no one had found anything paradoxical about it. But nothing could guarantee to Hilbert that his logic would not suffer the same fate as Frege's, sooner or later: someone might one day prove a thing and its opposite, which would deal the deathblow to predicate logic. This was the real reason why Hilbert strove to replace reasoning by a computing operation that would indicate whether a proposition is true or false. If he succeeded, the algorithm could not yield two different results; by construction, the logic thus achieved would necessarily be non-contradictory.

THE ELIMINATION OF THE INFINITE

The transition from computing to reasoning, which took place in Greece in the fifth century B.C., was caused by the irruption of the infinite into the field of mathematics. How can we hope to return to computation, when the infinite is everywhere in mathematics? To

answer this question, look at the example of a well-known algorithm
for deciding whether a polynomial equation such as $x^3 - 2 = 0$ or
$x^3 - 8 = 0$ has a solution in the integers. This algorithm can be
viewed as a decision algorithm for propositions of the form "there
exists x such that $P(x) = 0$" where P is a polynomial. First, if we look
for a solution between 0 and 10, a simple algorithm is to try each
number between 0 and 10, one after another. For instance, with the
polynomial $x^3 - 2$, none of the results of this process $-2, -1, 6, 25$,
$62, 123, 214, 341, 510, 727$, and 998 equals zero, so the equation
$x^3 - 2 = 0$ has no solution among the integers between 0 and 10.
This simple method stopped working in the fifth century B.C., when
mathematicians started looking for solutions in the infinity of all
integers and had to replace computing with reasoning.

In this particular case, however, we shouldn't jump to conclusions. When x is bigger than 10, x^3 is bigger than 1,000 – how could
it equal either 2 or 8? If one of these two equations has a solution,
it is necessarily smaller than 10, in which case it suffices to review
all potential solutions from 0 to 10. More generally, with any given
polynomial, we can calculate a number beyond which the highest
degree term will be too big to be offset by the other terms – in other
words, a number beyond which there can be no solution. By reviewing all the numbers smaller than this bound, we can answer the
question, that is, determine whether or not a solution exists in the
infinity of integers.

There is no reason to be awed by the quantifier "there exists" in
the proposition "there exists x such that $x^3 - 2 = 0$," even though
it evokes the infinity of integers all at once. In some cases, we can
eliminate this quantifier as the proposition is replaced by the equivalent "there exists x between 0 and 10 such that $x^3 - 2 = 0$," whose
truth or falsity can be determined by computation. This method
is called the "elimination of quantifiers," and it was precisely what
Presburger, Skolem, and Tarski used to build their algorithms.
Neither Presburger, nor Skolem, nor Tarski had succeeded in showing that quantifiers could be eliminated when one allowed both
addition and multiplication in number theory, but in Hilbert's day,
there was no evidence to show it was impossible. And, beyond
arithmetic, could a similar method be found that encompassed the
whole of mathematics?

CHURCH'S THEOREM

In 1936, the solution to Hilbert's decision problem was reached, independently, by Alonzo Church and Alan Turing, and their answer was negative: there is no decision algorithm for predicate logic. So there is a difference in kind between reasoning and computation, and Hilbert's program to replace reasoning with computation was doomed.

How did Church and Turing prove this? We observed that in the twentieth century, reasoning became itself an object of study. For this to happen, it had been necessary to give an explicit definition of the rules of inference and of the grammar of propositions used in reasoning processes. Similarly, to solve the decision problem, Church and Turing had to make computation the object of study. It was not enough to propose algorithms, as Euclid and the medieval mathematicians had been content to do; to prove that no algorithm exists for solving certain types of problem, they needed an explicit definition of the notions of algorithm and computable function. Mathematicians of the thirties offered several definitions: Jacques Herbrand and Kurt Gödel made one proposal (Herbrand–Gödel equations), Alonzo Church another (Lambda calculus), Alan Turing a third (Turing's machines), Stephen Cole Kleene a fourth (recursive functions), and there were others.

All these definitions turned out to be equivalent, and more or less, they all describe computation as a series of processing stages. When we calculate the greatest common divisor of 90 and 21, for instance, we carry out successive transformation operations: first, we turn the pair $(90, 21)$ into the pair $(21, 6)$, then transform this pair into the pair $(6, 3)$, and finally tranform this pair into the number 3. If we use $gcd(90, 21)$ to denote the greatest common divisor of 90 and 21, then the expression $gcd(90, 21)$ turns into $gcd(21, 6)$, then $gcd(6, 3)$, and finally 3. We say that the expression $gcd(90, 21)$ is "rewritten" successively as $gcd(21, 6)$, $gcd(6, 3)$, and 3. This notion of transformation, or rewriting, is common to the definitions offered by Herbrand and Gödel, Church, Turing, Kleene, and others. To this day, it is lodged at the heart of the theory of computation.

Euclid's algorithm, for instance, is made up of two rules of computation. The first transforms the expression $gcd(a, b)$ into $gcd(b, r)$

when the division of a by b does not work out exactly and where r is the remainder; the second transforms the expression $gcd(a, b)$ into b when the division of a by b does work out exactly.

Computation is thus defined as the gradual transformation of one expression into another, the process being guided by a body of rules. Interestingly enough, this resembles the definition of reasoning: inference rules, too, allow one expression to be replaced by another. For instance, one rule allows us to deduce the proposition B from the propositions "if A then B" and A. Can this rule be viewed as a computation rule transforming one problem (proving proposition B) into another (proving propositions "if A then B" and A)? What, fundamentally, is the difference between a computation rule and an inference rule?

Mathematicians of the early 1930s clearly identified that difference: when you transform the expression $gcd(90, 21)$ into $gcd(21, 6)$, then $gcd(6, 3)$, and then 3, you know from the start of the transformation process that, after a few stages, you will reach a result and the operation will come to a halt. When you use the rule that allows you to transform B into "if A then B" and A, the transformation process can be infinite. The proposition B turns into the proposition "if A then B" and A, the proposition A can in turn be similarly transformed, and so on, and so on. As a result, although this transformation process succeeds when the proposition is provable, it doesn't fail when the proposition is not provable but keeps searching endlessly for a solution.

For a body of computation rules to define an algorithm, it is necessary that these rules have an extra property ensuring that they will always reach a result after a finite number of stages: this property is called halting, or termination. It captures the difference between the general notion of computation method (defined by some body of computation rules) and the notion of algorithm (defined by a body of computation rules that necessarily halt at some point). According to this distinction, Euclid's algorithm is, indeed, an algorithm. The method defined by the rule "$gcd(a, b)$ turns into $gcd(a + b, b)$," however, is not, because the expression $gcd(90, 21)$ would be transformed into $gcd(111, 21)$, then $(132, 21)$, ... and so on, without ever reaching a result.

Hilbert's decision problem was not whether reasoning could be replaced by a computation method, because the answer to that is obviously yes: all you need to do is replace each inference rule with computation rule. The question was whether reasoning could be replaced by an algorithm, in other words, with a process that always halts and that says no when a proposition is not provable.

ALGORITHMS AS OBJECTS OF COMPUTATION

Euclid's algorithm and the addition algorithm apply to numbers. An algorithm can also be applied to other types of data – as in calculus, where algorithms are applied to functional expressions. Nothing prevents us from conceiving algorithms and computation methods that will be applied to bodies of computation rules.

One such method is called an "interpreter": it is applied to two objects a and b, where a is a set of computation rules, and it calculates the result that would be reached if the method formed by the rules of a were applied to b. We'll illustrate this abstract and somewhat obscure definition with an example. We have seen that Euclid's algorithm relies on two rules. Call a the set containing these two rules and b the pair of numbers 90 and 21. When we apply the interpreter U to a and b, we obtain the result 3, which we also would reach by applying Euclid's algorithm to the pair b. Of course, when the application of the method formed of the rules doesn't terminate, the application of the interpreter U to a and b doesn't terminate either. Nowadays, this notion of interpreter is a basic tool in the theory of programming languages: when a new programming language is invented, there is a priori no computer capable of executing the programs written in this language. Before the language can be used, an interpreter must be written in an already existing language. This interpreter is then applied to a program a written in the new language, and it calculates the result of that program when applied to a value b.

THE HALTING PROBLEM

In 1936, another attempt to design an algorithm that applies to rules of computation led to the first negative result of the theory of computability, a result that proved for the first time that certain problems cannot be solved by computation. The goal was to devise an algorithm *A* that, like the interpreter, would be applied to two objects *a* and *b*, where *a* is a set of computation rules, and would indicate whether the method *a* terminates or not when applied to the value *b*. If it does terminate, applying *A* to *a* and *b* returns 1; otherwise it returns 0.

It turns out that no such algorithm exists. In order to prove this, in 1936 Turing and, independently, Church and Kleene built a proof by reductio ad absurdum: they assumed the existence of the hypothetical algorithm and demonstrated that its existence has contradictory consequences – and, therefore, that the hypothesis had to be false. Let's assume that *A* exists. The trick is to construct a method *B* that, given *a* and *b*, embarks on some computation that terminates when *A* returns 0 and doesn't terminate otherwise. Thus we see that *B* applied to *a* and *b* terminates precisely when *a* applied to *b* does not. Next, we construct a method *C* that, given *a*, applies *B* to *a* and *a*. We finally ask: what happens when *C* is applied to itself? Does it terminate or not?

A moment's thought reveals that *C* applied to *C* terminates, by definition, exactly when it does not – which is absurd. This contradiction shows that the method *C* cannot exist, proving that the method *B* cannot exist either, and hence ruling out the existence of the algorithm *A*. This theorem, that the halting problem cannot be solved by an algorithm, is called the "theorem of the undecidability of the halting problem."

Applying this theorem to Hilbert's decision problem is, actually, the simplest step in the process, as Church and Turing understood, again independently. Predicate logic was designed to permit the expression of all forms of mathematics, so it allows us to express propositions of the form "the method defined by the rules *a* terminates when applied to the value *b*." Once again, we'll deploy reductio ad absurdum. If there were an algorithm to determine whether or not a proposition is provable in predicate logic, it could determine, more particularly, whether or not a proposition of this form is provable, and therefore would be able to determine whether or not the algorithm *a* terminates when applied to the value *b* – contradicting the undecidability of the halting problem. This result – that there is no algorithm to decide whether or not a proposition is provable in predicate logic – although obtained independently by both Turing and Church is called "Church's theorem."

So computation and reasoning are two different things indeed. Some mathematical problems cannot be solved by computation and require reasoning. They justify the transition from the prehistory of mathematics, when algorithms ruled supreme, to the mathematics of the ancient Greeks.

ANALYTIC DOES NOT MEAN OBVIOUS

Church's theorem sheds new light on the notion of analytic judgment and particularly on its relationship with the notion of evidence. Traditional examples of analytic judgments, or tautologies, often have an air of the obvious: triangles have three angles, herbivores eat grass, a promise is a promise, and so on. So much so that in everyday life "tautology" is synonymous with "truism." Similarly, saying that something is true "by definition" is often of a way of saying that it is obvious.

It is therefore understandable that so many mathematicians were reluctant when Frege (and later others) developed the thesis according to which all mathematical judgments are analytic.

More generally, it is an old criticism that logical reasoning does not produce anything new – all it does is make explicit what already lies implicit in the axioms. For example, the syllogism "all men are mortal, Socrates is a man, therefore Socrates is mortal" does not seem to hold any major revelation – Socrates being a man, the conclusion that he is also mortal is implicit in the premise "all men are mortal."

Church's theorem sweeps these arguments aside: certainly, logical reasoning reveals only those truths already implicit in the axioms but, contrary to the impression we have of the syllogism "all men are mortal, Socrates is a man, therefore Socrates is mortal," these truths, although implicit in the axioms, are far from being obvious or truisms. If they were, there would exist an algorithm for finding out what conclusions can be drawn from a set of axioms, and Church's theorem shows the exact opposite. This means that the operation of making explicit what was implicit – namely, reasoning – is far from insignificant. Metaphorically, this operation may be compared to the work of a prospector, panning for gold in a river, sifting the sand for months in search of a nugget. Of course, the nugget was in the sand from the very start, but it would be an overstatement to claim that finding it is as trivial as bending over and picking it up. Even when a result is implicit in the axioms, making it explicit yields information and knowledge.

This distinction between "analytic" and "obvious" can be refined by taking into account the amount of computation needed to reap results. Even in cases when there is an algorithm that answers a question, if this algorithm demands lengthy calculations, we can consider the answer it produces as not obvious. For example, many algorithms exist for determining whether a number is prime or composite; however, determining the primality of a very large number may take years of computation. Despite the existence of the necessary algorithms, we cannot say that the primality of a number is obvious. Determining a number's parity is easy – we just look at its last digit – but determining a number's primality certainly is not.

This mistake of confusing "computable" with "analytic" is common in discussions about the mathematization of the natural sciences. We frequently hear that, since mechanics was mathematized – that is, transformed into an axiomatic theory – in the seventeenth century, solving a mechanics problem (for example, predicting the position of a planet at a future date) requires nothing but simple computation. This argument is in conflict with Church's theorem: now that mechanics is mathematized, solving a problem requires reasoning, not computation.

It remains to be seen whether mechanics can be expressed by a set of algorithms, instead of a set of axioms – but more on that later.

CHAPTER FIVE

Church's Thesis

THE ATTEMPT to address Hilbert's decision problem and the negative result it achieved in Church's theorem led mathematicians of the thirties to clarify what an algorithm is. They offered many definitions, among which were the Herbrand–Gödel equations, Church's lambda calculus, Turing machines, Kleene's recursive functions, and rewrite rules. Each of these puts forward a language in which to express algorithms – nowadays, we would say that each defines a "programming language."

Time has shown these definitions to be equivalent: if an algorithm can be defined in one of these programming languages, that definition can be translated into any of the others. This equivalence of definitions ranks among the greatest successes of computability theory: it means that an absolute concept of computation has been reached independent of the accidental form assumed by this or that algorithmic language.

Yet the mathematicians of the thirties faced an obvious question. Is this it? Is this "the" notion of computation? Or might someone in the future come up with other languages capable of expressing more algorithms? Most mathematicians of the thirties rejected this possibility; to them, the concept of computation, defined by Turing machines or by lambda calculus, was the right one. This thesis is called Church's thesis, although as before several mathematicians – particularly Turing – developed similar ideas.

THE COMMON CONCEPT OF COMPUTATION

Church's thesis asserts the identity of two concepts: the concept of computation as defined by lambda calculus, Turing machines, and so forth and the "common" concept of computation. One of the reasons Church's thesis is so hard to formulate precisely is that it is unclear what the common concept of computation is. Indeed, if you formulate Church's thesis by stating that no algorithmic language that might be put forward in the future will ever be more powerful than those we already know or, in other words, that all the algorithms we may come across in the future will be expressible in the languages we use today, you sound more like a fortune-teller looking into a crystal ball than a mathematician formulating a scientific theory. In the thirties, mathematicians therefore tried to refine this "common" concept of an algorithm in order to give Church's thesis a more precise statement.

There are two ways of achieving this, depending on whether the computing task is to be carried out by a mathematician or by a physical system, that is, a machine. This prompts us to distinguish between two variants of Church's thesis: its psychological form and its physical form. According to the psychological form of Church's thesis, all the algorithms that a human being is capable of executing in order to solve a specific problem can be expressed by a set of computation rules. The physical form of Church's thesis, on the other hand, states that all the algorithms that a physical system – a machine – is capable of executing systematically in order to solve a specific problem can be expressed by a set of computation rules.

If we adopt the materialist point of view, according to which human beings are part of nature and thus have no supernatural faculties, then the psychological form of Church's thesis is a consequence of the physical form.

Yet it is important to note that, even in that case, the two theses are not equivalent. The physical thesis is by no means a consequence of the psychological one. Nature might very well be capable of computing more things than a human being can. Some physical systems – pocket calculators, for instance – can compute more things than, say, a sparrow, and there is no a priori argument that the same does not apply to human beings. However, if the physical

form of Church's thesis is true – in other words, if all the algorithms that a physical system is capable of computing really can
be expressed by a set of computation rules – then whatever nature
can compute, the human being can compute, too. This is tantamount to saying that the human being is among the best computers in nature: there is nothing in the whole natural order, animal or
machine, that can compute better than a human being. This thesis, which we will call the "computational completeness of human
beings," can be seen, in a way, as the converse of materialism. It
can be defended independent of Church's thesis. Likewise, assuming that the psychological form of Church's thesis and the thesis of
the computational completeness of human beings are true, then the
physical form of Church's thesis is true as well. Indeed, everything
that nature is capable of computing can be computed by a human
being, and everything that a human being is capable of computing
can be expressed by a set of computation rules.

A little formalization can help to clarify how these different theses
relate to one another. We are dealing with three sets of algorithms:
the set R comprising algorithms expressible by a body of computation rules, the set M of algorithms that can be computed by a physical system, and finally the set H of algorithms that can be computed
by a human being.

We can now write down six relations, or "theses," of the form
$A \subset B$, meaning that A is a subset of B. Two of them certainly hold,
namely $R \subset M$ and $R \subset H$. The other four are:

- The physical form of Church's thesis, $M \subset R$;
- The psychological form of Church's thesis, $H \subset R$;
- The materialist thesis, $H \subset M$; and
- The thesis of the computational completeness of human beings,
  $M \subset H$.

Using the fact that, when A is a subset of B and B is a subset of C,
then A is a subset of C, we draw four conclusions:

- If the physical form of Church's thesis and the materialist thesis $H \subset M$ are true, then so is the psychological form of Church's
  thesis;

- If the physical form of Church's thesis is true, then so is the thesis of the computational completeness of human beings;
- If the psychological form of Church's thesis and the thesis of the computational completeness of human beings are true, then so is the physical form of Church's thesis;
- If the psychological form of Church's thesis is true, then so is the materialist thesis.

THE PHYSICAL FORM OF CHURCH'S THESIS

Neither form of Church's thesis can be proved by purely mathematical means, because they both draw on concepts external to mathematics – one relies on the concept of human being, while the other relies on the concept of physical system. Only by appealing to principles of psychology or physics can we argue for or against these two theses.

In 1978, Robin Gandy proposed a proof of the physical form of Church's thesis. This proof is based on the assumption that physical space is the ordinary Euclidean three-dimensional space. Further, the density of information as well as the speed at which information can be transmitted are both assumed to be finite. The first hypothesis means that a finite physical system can take on only a finite number of different states. The second hypothesis means that the state of one system can influence that of another system only after a certain delay, which is proportional to the distance between the two systems.

Gandy's proof concerns a physical system that we choose to observe at successive instants, say every second or every microsecond. If we assume, to start with, that the size of this system is finite, then the first hypothesis implies that it can take on only a finite number of states and, moreover, its state at a given instant depends solely on its state at the previous instant. From this, we deduce that the state of the system at any given time can be computed from its initial state by means of a set of computation rules.

However, assuming a priori that the system is of finite size is not satisfactory. When you compute the product of two numbers, you use a sheet of paper of finite dimensions, but the multiplication algorithm is not limited to the numbers written on that sheet.

Therefore, to define the physical notion of computing, we must consider systems of infinite size.

Gandy therefore moved on to suggest that we split this infinite system into an infinite set of identical, finite cells. According to the first hypothesis, at any given moment, each cell can take on only a finite number of states. According to the second hypothesis, the state of a cell at any given moment depends only on the states of this cell and a finite number of neighboring cells at the previous moment.

At the beginning of the computing process, all but a finite number of cells are in a quiescent state. Then the system evolves step by step. After a predefined number of steps, or when one of the cells reaches a certain state, the computing process comes to an end and its result is the state of the whole system. Gandy has thus shown that the state of the system at any given time can be calculated from its initial state thanks to a set of computation rules and that, as a consequence, everything that can be computed by this physical system can just as well be computed by a set of computation rules.

Gandy's argument, as presented here, applies only to deterministic systems whose evolutions are wholly determined by their initial states. The argument can easily be extended to nondeterministic systems, however: all you need to do is replace actual states of the system with sets of possible states.

It is important to note that Gandy's argument does not assume that nature is discrete, that is, split into cells. The division of space and time into cells is a mere methodological artifice. The finiteness of the number of states that each cell can assume and of the number of cells that can influence the next state of a given cell is just a consequence of the hypotheses that the density of information and its transmission speed are both finite. Likewise, Gandy's proof by no means reduces nature to a computer – such a standpoint would be the present-day equivalent of the mechanistic theory that, in the eighteenth century, reduced nature to a giant piece of clockwork. However, even without adopting a point of view of this sort, one may raise, among many questions about nature, that of the computations that can be executed by a physical system.

On the other hand, it is not unreasonable to criticize Gandy's hypotheses. They allow a classic Euclidean space to which is added two principles taken from modern physics – finite density of information and finite information transmission speed. Does Church's thesis remain valid when other hypotheses about nature are formed, by quantum physics, for example, or by the relativistic theory of gravitation? Even in a world where new physics theories flower every day, in which it is difficult to have any complete certainty, we have reason to think it does. By arguments close to those suggested by Gandy it looks likely that Church's thesis is valid in all these new theories. None of these theories seems to suppose the existence of a physical system able to compute more things than a set of computation rules can.

MATHEMATIZING NATURE

Church's thesis sheds new light on a old philosophical problem: are mathematical concepts fit for describing nature?

Let's illustrate this problem with a classic example. When Johannes Kepler, in the seventeenth century, following in the footsteps of Tycho Brahe, observed the motion of planets around the sun, he realized that their trajectories were ellipses. Ellipses, however, were not invented in the seventeenth century to better describe the motion of planets – they were already known in the fourth century B.C. So why did this concept, invented in antiquity for reasons that had nothing to do with celestial mechanics, turn out to be so well-suited to describing the trajectories of planets? More generally, why do planets move along simple geometric figures or, in other words, why can their movements be described in mathematical terms? Albert Einstein summed up this astonishing fact by observing that the most incomprehensible thing about the world is that it is comprehensible.

Until the sixteenth century, one could simply brush this question aside as, back in the days of Aristotelianism, nature was not believed to obey mathematical laws. Since then, however, mathematics and physics have come a long way and one is bound to acknowledge that, as Galileo put it, the great book of nature is indeed written in the language of mathematics.

There have been several attempts to explain why the laws of nature can be mathematized so well. One hypothesis suggests that nature was created by a mathematician-God who chose to write the great book of nature in the language of mathematics and decided that the planets would describe ellipses. This explanation solves the problem but is hard to verify, and there is no good reason to give it any preference. Besides, it only replaces one mystery with another – why should God be a mathematician?

Another explanation supposes that people developed, or rather abstracted, mathematical concepts from their observation of nature. It should therefore come as no surprise that objects in nature show similarities to mathematical objects. Yet this argument, which may explain why some mathematical concepts describe natural phenomena so closely, does not suffice when it comes to the concept of ellipse and the motion of planets, because the former was not born of observing the latter.

A variant suggests that scientists observe only carefully chosen phenomena, namely the things they can mathematize, and overlook the others. For example, the motion of two bodies – say, the sun and a planet – can easily be described in mathematical terms; yet, as the work of Poincaré showed in the early twentieth century, describing the motion of three bodies is much more difficult. By some stroke of good fortune, it happened that physicists of the seventeenth century focused their attention mainly on the motion of two bodies and somewhat neglected that of three bodies. This argument has some relevance but it is not sufficient. It leaves open the question of why certain phenomena – even if not all – can be mathematized so readily.

One may be tempted to suspect scientists of simplifying phenomena so as to make them mathematizable. For instance, due to their mutual attraction, planets actually describe approximate ellipses, yet this fact is often overlooked – not only because the mutual attraction of planets is weak, but also because the simplification is a welcome one. There is some truth in this explanation, but one still wonders why natural phenomena are so easily approximated – if not described exactly – by mathematical concepts.

The last explanation draws on the materialist hypothesis, according to which we are ourselves part of nature and our mathematical concepts are therefore fit to describe it. This explanation depends on the troublesome premise that a mechanism is better understood from the inside than from the outside. Experience seems to teach us otherwise. We understand, nowadays, how the human liver works, not because we each have a liver and have found the key to its functioning by dint of introspection, but because experimenters have observed other people's livers – from the outside.

In some cases, the fact that a phenomenon appears to obey a mathematical rule explains itself once the phenomenon is better understood. Recall the explorer, whom we left reflecting on the origin of the word "fish," and imagine that he goes continues to explore his island and comes across an unknown species of fruit tree. He picks some fruit, weighs it, and realizes that the fruit have very different masses, so he decides to use the mass of the lightest fruit as a unit. He then notices, much to his surprise, that the different masses are always whole multiples of this unit. There are fruits of mass 1, fruits of mass 12, others of mass 16. But he cannot find a fruit of mass 45. Yet he is convinced that such a fruit exists; in fact, he foretells its existence, so he carries on looking and eventually finds one. It seems amazing that nature should obey a rule of arithmetic so strictly, so much so that those rare observations that seem to depart from the rule are, in fact, incomplete. But, one day, the explorer manages to slice the fruit in half and realizes that its mass is mostly due to the seeds that make up its core. These seeds are all identical and have the same mass – a fruit of mass 12 is simply a fruit containing twelve seeds; the explorer just hadn't happened to come across a fruit containing 45 seeds during his first series of observations. The flabbergasting mathematical regularity of the fruits' masses, which the explorer had attributed to a mathematician-God who created the fruit to conform to a chosen arithmetic structure, turned out to have a much more prosaic explanation. Of course, this discovery leaves some questions unanswered – why the seeds are all of identical mass – but the mystery of the arithmetic regularity is solved.

Something similar occurred in the nineteenth century when chemists, among them Dimitri Mendeleev, noticed that the atomic masses of chemical elements displayed a regular arithmetic structure – except for the atomic masses of 45, 68, and 70, which were missing. These gaps led Mendeleev to foretell the existence of three chemical elements, all of which were later discovered: scandium, gallium, and germanium. Understandably, this arithmetic regularity astonished people. Yet the fact seemed less of a miracle once scientists realized that the mass of atoms is mostly due to the particles that make up their nuclei and that have roughly the same mass, and once it became clear that Mendeleev had merely foretold the possibility of an atom's having a nucleus made up of 45 particles. In the near or distant future, the seemingly miraculous arithmetic regularity of the elementary particles may receive a similar explanation based on the decomposition of these particles into smaller entities.

By contrast, it seems impossible to apply this kind of argument to regular, geometric motion of planets, for instance. There appear to be several sorts of mathematical regularity of phenomena and, rather than search for a global explanation, it may be wiser to seek one explanation for each type of regularity – the key to mathematical regularity may be different in chemistry than in celestial mechanics.

Let's focus on a single phenomenon, which is not unrelated to the figures described by planets: the law of gravity. Suppose you drop a ball from the top of a tower and measure the distance it covers in one, two, three seconds. Suppose you manage to create a vacuum around the tower. Observation of this phenomenon reveals that it follows a simple mathematical law: the distance covered is proportional to the square of the time elapsed. This law is expressed by the following proposition: $d = 1/2 \, gt^2$ (with $g = 9.81 \mathrm{ms}^{-2}$).

Several mysteries cry out for explanation. One is that, if you repeat the experiment, you get an identical result: the distance covered during the first second of free-fall remains the same. It is not clear why gravity is a deterministic phenomenon, but Church's thesis doesn't have anything to say here. Another mystery is that the relationship between time and distance can be expressed by a mathematical proposition: $d = 1/2 \, gt^2$. Church's thesis may well have something to say about that.

The physical system comprising the tower, the ball, the clock used to measure the time, and the height gauge used to measure the distance is a calculating machine. When you pick a duration, say 4 seconds, let the ball freefall during that time, and measure the distance covered (78.48 m), you carry out a computing task using this machine, which maps a value (4) to another one (78.48). According to Church's thesis, the computing task performed by this special calculating machine might have been performed just as well by a set of computation rules. In that case, the algorithm that gives the same result as this analogical machine consists in squaring a number, multiplying it by 9.81, and then dividing it by two. When you express this algorithm in the language of computation rules, you give it a mathematical form. The physical Church thesis implies that the law of gravity can be expressed in mathematical language.

The mystery surrounding this thesis has thickened. At first glance, it seemed to imply that the language of computation rules, in any of several equivalent forms, is powerful enough to express all possible algorithms. In fact, as David Deutsch underlined, the physical form of Church's thesis also expresses a property of nature, and one consequence of this property is that natural laws are captured by this notion of computation rule, which explains why they are mathematizable.

We must now look with new eyes at Gandy's argument in defence of Church's thesis. Gandy claimed that Church's thesis is true because the density and transmission speed of information are finite. If you piece these arguments together, you reach the following conclusion: because the density and transmission speed of information are finite, the law of gravity can be described in mathematical language.

Likewise, it appears that all natural laws are mathematizable because the density and transmission speed of information are finite. Yet the physical form of Church's thesis seems to strengthen this reasoning: if physicists were to give up these hypotheses, and if other properties of nature were used to confirm Church's thesis, this explanation of nature's mathematizability would still be valid.

Thus the physical form of Church's thesis alone implies that we can mathematize the laws of nature. Alternatively, this conclusion follows from the psychological form of Church's thesis together with the computational completeness of human beings.

We can, in fact, argue this point more directly, going back to the example of the falling ball. If human beings are the best computers in the whole natural order, then a human being is capable of computing the distance covered by the ball during a given time, because nature is capable of doing so. Now, according to the psychological form of Church's thesis, the algorithm that yields the distance covered by the ball during the duration of its fall must be express-ible by a set of computation rules, and therefore by a mathematical proposition.

This provides a bridge to an argument we offered earlier – that we human beings are part of nature, therefore our mathematics is fit to describe nature. Instead of naively supposing that being part of a system is enough to understand it, from within, this new explana-tion merely relates our computing capacity to that of nature. From this relationship, we may deduce not laws of nature but the poten-tiality of their description – a potentiality that has yet to be realized.

The idea that the mathematizability of natural laws is a conse-quence of Church's thesis has been in the air for about fifteen years. Related theses can be found in the writings of David Deutsch or John Barrow, for instance. However, both Deutsch and Barrow combine Church's thesis with other notions – universality, complexity, and so on – that seem unnecessary to this author.

Many points still need clarifying in this rough outline of an expla-nation. First, even if the law of gravity can be expressed in a proposi-tion, nothing accounts for the proposition's baffling simplicity. The arguments claiming scientists select mathematizable phenomena or resort to approximation may come into play in this context, as long as the word "mathematizable" is replaced by the word "sim-ple." This then leaves us to characterize the type of mathematical regularities that can thus be explained.

At the least, this sketch of an explanation has the merit of relating the a priori purely epistemological question of nature's mathema-tizability with some objective properties of nature – for example, the finiteness of the density and transmission speed of information, Church's thesis, and so forth. From a methodological point of view, this explanation can be credited with showing that, in order to understand why nature is mathematizable or, in other words, why its laws are expressible in a language, we may need to focus not only on the properties of language but also on the properties of nature.

THE FORM OF NATURAL LAWS

Another virtue of this explanation is that it leads us to reflect on the mathematical form of theories about nature. Galileo's observation that the great book of nature is written in the language of mathematics led physicists to describe the laws of nature using mathematical propositions. For instance, the distance covered by a free-falling ball and the time it takes for the ball to cover that distance are connected by the proposition: $d = 1/2 \, gt^2$.

What still wants explaining, after Galileo, is why such a proposition exists. If you accept the theory that this proposition holds because there exists an algorithm that enables you to calculate d when you know the value of t, your can take as your objective to connect these physical quantities not by a proposition but by an algorithm. Admittedly, in this example, it is trivial to compute d from t. If nature, however, is not only mathematizable but also computable, nothing justifies our limiting ourselves to expressing its laws by propositions: we can dare to aspire to expressing them by algorithms.

Actually, this algorithmic reformulation of the natural sciences has already been achieved in at least one field: grammar. In order to understand why a language's grammar is a natural science just like physics or biology, you need to put yourself in the explorer's shoes again. After completing his botanical experiments, he stumbles across a people speaking an unknown language and sets about describing that language. Just like a physicist or a biologist, the explorer is confronted with facts, which, in this particular case, are enunciations, and he must develop a theory to account for them, that is, to explain why this sentence can be enunciated but that other one cannot. This theory is what we call "grammar." An English grammar, for instance, must explain why "still waters run deep" can be enunciated, while "run still waters deep" cannot.

Traditionally, a grammar is expressed by a body of propositions called "grammar rules" – thus, in the English grammar, "the adjective never changes form" or, in the French grammar, "the adjective agrees with the noun." These rules enable us to deduce that the sentence "still waters run deep" is properly formed, whereas the sentence "stills waters run deep" is not.

Enunciations take place in nature and Church's thesis tells us that there exists an algorithm, or at least a computing method, that indicates whether or not a sentence is correct in a given language. From the materialist's standpoint, the fact that human beings are involved in the enunciating process has no effect on the fact that these phenomena must abide by the same rules as other natural phenomena.

In this specific case, however, there is no need to call on any high principle. One thing is certain: for speakers to use the language, they need to be able to decide whether a sentence is properly formed or not. According to the psychological form of Church's thesis, there must therefore exist an algorithm or at least a rule of thumb allowing the speaker to know whether a sentence is or is not properly formed. Hence, a grammar must be expressible not only as a body of propositions but also as an algorithm.

In the late 1950s, Noam Chomsky suggested that we express the grammar of natural languages by means of algorithmic forms. In an effort to demonstrate this, he developed a way of formulating algorithms that enabled him to write a grammar in an algorithmic form without straying too far from the traditional form of grammars, usually presented as sets of rules.

In devising an algorithm that will indicate whether or not a sentence is properly formed, for example in English, you face several problems that were previously screened by the traditional form of grammars. Thus, in order to make sure that the adjectives do not change form in the sentence "still waters run deep" you need first to find the adjectives. This leads us to wonder how we know that "still" is an adjective qualifying the noun "waters," or how an explorer, happening upon *us* and desirous of learning our language, might spot the adjectives in a sentence. Do we know that "still" is an adjective because the dictionary tells us so? Is it because we know the meaning of the words "still" and "waters"? Is it because the word "still" is placed before the word "waters"? Or is it precisely because it does not change form? Reformulating the grammars of natural languages has brought many new questions to the fields of grammar and linguistics, not only about languages, but also about the way our brains use them.

The example of grammar is not an isolated case. Since the 1980s, computer scientists have been taking a new interest in quantum physics and more specifically in the principle of superposition – the object of "quantum computing" is to try to exploit this principle in order to execute several computing tasks simultaneously on a single circuit, potentially leading to a revolution in computing power.

They are also turning to biology and more precisely to the functioning of the cell, which could suggest original computing processes. This effort is called "bioinformatics."

A second and less publicized part of these projects aims at reformulating parts of physics and biology into algorithmic form. In time, they may transform not only computer science but also the natural sciences.

CHAPTER SIX

Lambda Calculus, or an Attempt to Reinstate Computation in the Realm of Mathematics

THE THEORY OF COMPUTABILITY, which provided a negative answer to Hilbert's decision problem, went hand in hand with an attempt to reinstate computation in mathematics. Although this attempt failed, it is significant and worthy of attention. In many ways, it fore-shadowed future developments in the history of mathematics. This attempt rests on a language that is considered to be Church's greatest achievement, namely lambda calculus.

Lambda calculus started out as a simple functional notation system. A function can be represented by a table of values, or by a curve, but also, in some cases, by a functional expression. For example, the function that, to a number, assigns the square of that number can be represented by the expression $x \times x$.

This notation, however, is somewhat clumsy, because it doesn't distinguish between the function and the value assumed by that function at a specific point. Thus, when you say that "if x is odd, then $x \times x$ is odd," you are talking about the number $x \times x$ or, in other words, about the value taken by the function at x. On the other hand, when you say that "$x \times x$ is increasing," you are talking about the function itself. In order to avoid this confusion, we will denote the function by $x \mapsto x \times x$ rather than $x \times x$.

This new notation is believed to have been introduced around 1930 by Nicolas Bourbaki, the pseudonym used by a group of mathematicians who wrote an important treatise, published in a series of installments over many years, that compiled the mathematical knowledge at the time. Around the same time, Church introduced a similar notation: $\lambda x(x \times x)$, where the arrow $\mapsto$ was replaced by the Greek letter *lambda*. This notation is reminiscent of a previous convention, used by Russell and Whitehead in the early 1900s: $\hat{x}(x \times x)$. As the story goes, Church's publisher did not know how to print a circumflex on top of an x, so instead he had the letter preceded by a symbol resembling a circumflex, namely the capital letter *lambda* (Λ), which was later replaced by the lowercase letter *lambda* (λ). Although Bourbaki's notation ($x \mapsto x \times x$) prevailed, Church's notation is still used in logic and computer science, and the very name of this new language is derived from Church's notation: it is called lambda calculus.

As we have seen, one of the most important operations in predicate logic is replacing a variable by an expression. For example, from the proposition "for all numbers x and y, there exists a number z that is the greatest common divisor of x and y," we can infer the proposition "there exists a number z that is the greatest common divisor of 90 and 21" by replacing the variables x and y with the numbers 90 and 21. In this example, x and y represent numbers, but variables can also stand for functions, in which case they are replaced with expressions such as $x \mapsto x \times x$. Russell and Whitehead had already raised this possibility: they wondered what is obtained by replacing the variable f, for example, with the expression $x \mapsto x \times x$ in the expression $f(4)$, and they reached the commonsense conclusion: 4×4.

In 1930, however, Church pointed out a minor drawback of that result: when we replace the variable f with the expression $x \mapsto x \times x$ in the expression $f(4)$, the result is not 4×4, as Russell and Whitehead suggested, but $(x \mapsto x \times x)(4)$. This expression can then be transformed into 4×4, but in order to do so, we need to call upon an additional rule of computation, namely the rule stating that the expression $(x \mapsto t)(u)$ turns into the expression t where the variable x is replaced by the expression u. According to Church, Russell and Whitehead had blended together two operations that it is important to separate: replacing the variable f on one hand, and evaluating the function, which leads to replacing the variable x, on the other. Church, who was apparently rather fond of the Greek alphabet, named "beta reduction" the computation rule that transforms the expression $(x \mapsto t)(u)$ into the expression t where the variable x is replaced with the expression u.

This computation rule may seem insignificant, but Church showed that, by applying it, you can simulate any computation operation you like. It may be unclear how the function that, to a pair of numbers, assigns the greatest common divisor of these numbers can be expressed in lambda calculus and computed with beta reduction, but it can. And what is true of this function is also true of all computable functions. Lambda calculus thus enables us to compute the same things as Turing machines or Herbrand–Gödel equations. Initially, Church even defined "computable" as "that which is expressible in lambda calculus." This definition of the notion of computability caused many a mathematician to raise a sceptical eyebrow, because they thought beta reduction too weak a rule to be able to compute everything. Only when the equivalence of lambda calculus and Turing machines was demonstrated did they acknowledge that Church's instincts had been right. The demonstration rests on the trick of applying a function to itself. Therein lies the unsuspected computational power of lambda calculus: this language allows a function to be applied to itself.

Church's most interesting idea, however, was yet to come. Unlike Turing machines, lambda calculus has the advantage of drawing on a traditional mathematical notion: that of function. As a consequence, it became possible to imagine a new formalization of mathematics – an alternative to predicate logic and set theory, based on the notion of function and on lambda calculus. Church put forward such a formalization in the early thirties. This theory relies, among other things, on beta reduction and, consequently, all functions can be expressed directly as algorithms. It thus reinstates computation as a central part of mathematical formalization.

Unfortunately, this theory is afflicted with the same flaw as Frege's logic: it is self-contradictory, as Stephen Cole Kleene and John Barkley Rosser proved in 1935. Church therefore abandoned the idea of using lambda calculus to formalize mathematics. Lambda calculus was allocated a less ambitious role as a language dedicated to the expression of algorithms.

Thereafter, Haskell, Curry, and Church himself all tried to restore the consistency of the theory. Curry strove to defend the idea according to which a function can be applied to itself, but the mathematical formalization he developed in the process strayed so far from common sense that he gave it up. Church, for his part, applied to lambda calculus the same method Russell had used on Frege's theory, ruling out the application of a function to itself. As a result, the theory was consistent again, but there was a price to pay: lambda calculus lost a great deal of its computational power, most of which stemmed precisely from the possibility of applying a function to itself. This version of the theory is called "Church's type theory" and is nowadays considered a variant of Russell and Whitehead's type theory and of set theory.

It is possible to formulate Church's type theory using a computation rule for beta reduction. However, because in the theory it is impossible to apply a function to itself, this computation rule is not nearly as powerful as in Church's original theory. So although Church's type theory grants a place to computation (because it makes room for this computation rule along with axioms), that place is modest.

Out of pique, perhaps, Church eventually transformed this computation rule into an axiom called the "beta-conversion axiom." The attempt to devise a formalization of mathematics that would make room for computation was over.

These three chapters dedicated to the theory of computability thus end on a paradox. The theory of computability assigns important roles to the notions of algorithm and computation; however, despite Hilbert's radical attempt to replace reasoning with computing, and despite Church's even bolder but unsuccessful attempt, the theory of computability did not revolutionize the notion of proof. All the time the theory of computability was being developed, proofs continued to be expressed in predicate logic and constructed with axioms and inference rules, in keeping with the axiomatic conception of mathematics – and, in those proofs, not the smallest place was left for computation.

CHAPTER SEVEN

Constructivity

WHEREAS THE THEORY OF COMPUTABILITY placed computation at the heart of several major mathematical issues, the theory of constructivity, which was developed independently, does not appear, at first glance, to assign such an essential role to computation. If you look closer, however, you will find that computation also plays an important part.

CONSTRUCTIVE VERSUS NONCONSTRUCTIVE ARGUMENTS

Let's start with a story. It takes place in Europe, shortly after World War I. The explorer has returned from his many travels and, now clean shaven and well rested, decides to ride the Orient Express from Paris to Constantinople. In his compartment, he finds an intriguing, scented note sent by a secret admirer: the mystery woman asks him to meet her on the platform of the last station the train calls at on French territory. The explorer gets hold of a list of the stations the Orient Express stops at from Paris to Constantinople: Strasbourg, Utopia, Munich, Vienna, Budapest, ... He knows that Strasbourg is in France and that Munich is in Germany, but he has no idea where Utopia is. He asks his fellow passengers, he asks the conductor, but Utopia is such a small, obscure station that nobody on board the train is able to help him – it is hard to fathom why the Orient Express would even stop in a town no one seems ever to have heard of! Will the explorer manage to meet his lady? It's not a sure thing. Yet it is not hard to prove that there exists a station at which the Orient Express stops that is in France and is such that the next station is no longer in France. For either Utopia is in France, and then is the train's last stop there, or it is not, in which case Strasbourg is the train's last stop on French soil.

Those who disdain train problems may prefer a more abstract formulation of this reasoning. If a set contains the number 1 but not the number 3, then there exists a number n such that the set contains the number n but not the number $n + 1$. Indeed, either the number 2 is contained by the set, and it is the solution, or it is not and the solution is 1.

This reasoning, however, doesn't help the explorer, who does not just want to know whether there exists a station where the mystery woman awaits him, but also which one it is. What good is mathematics if it can't get him to his rendezvous?

That night, on board the Orient Express, the explorer becomes aware of an idea that also hit the mathematicians of the early twentieth century: some proofs, such as the proof we have just outlined, demonstrate the existence of an object satisfying a certain property without specifying what that object is. These proofs must be distinguished from those that demonstrate the existence of an object satisfying a certain property *and* give an example of such an object. For instance, we can prove that there exists an Austrian city in which the Orient Express stops: Vienna. The kind of proof of existence that points out the object sought are called constructive proofs, and the object is called a witness of that existence. Conversely, the proof used by the explorer is called a nonconstructive proof.

When you think about it, the fact that it is possible to build nonconstructive proofs is surprising. Indeed, among inference rules, only one allows us to prove propositions of the form "there exists x such that A": this rule is called the "existential quantification introduction rule" and it allows us to infer the proposition "there exists x such that A" from an instance of A or, in other words, from a proposition similar to A in which the variable x has been replaced by some expression. For example, this rule enables us to infer the proposition "there exists an Austrian city at which the Orient Express stops" from the proposition "Vienna is an Austrian city at which the Orient Express stops." Whenever this rule is applied, the witness can be found in the instance of A that is used.

How do we lose the witness in course of a nonconstructive proof? Let's look again at the explorer's reasoning process. First, he demonstrates that, if Utopia is in France, there exists a last French station before the border. This part of his reasoning process is constructive. He first proves that Utopia is the last station before the border, then uses the existential quantification introduction rule: the witness is Utopia. Then he proves that, if Utopia is not in France, there still exists a last French station before the border – this part of the reasoning process is constructive as well. He first demonstrates that Strasbourg is the last station before the border, then uses the existential quantification introduction rule: the witness is Strasbourg. Finally, we come to the third part of his demonstration, in which he glues these two halves reasoning by cases: either Utopia is in France, or it is not. In both cases, there exists a last French station before the border. This is where the witness is lost, because although the two halves of the demonstration each produce a witness, Utopia or Strasbourg, it is impossible to choose between them.

"Reasoning by cases" is an inference rule that allows us to deduce the proposition C from three hypotheses of the forms "A or B," "if A then C," and "if B then C." In our example, propositions C, A, and B can be, respectively, "there exists a last station before the border," "Utopia is in France," and "Utopia is not in France." The first two parts of the explorer's reasoning process proved the propositions "if A then C" and "if B then C," but what proves the first proposition, "Utopia is either in France or not in France"? How do we know that Utopia is either in France or not? Here, we must resort to another inference rule: the excluded middle.

The principle of excluded middle is an inference rule that allows us to prove a proposition of the form "A or not A" without having to prove a premise. This rule expresses the common idea that if proposition "A" is not true, its negation "not A" is.

This principle seems obvious and yet enables very abstract reasoning processes. Imagine that we are sailing in a boat in the middle of the ocean and somebody throws a coin overboard. The coin sinks and drops to the bottom of the ocean, thousands of meters below. We have no way of knowing which side it landed on. The principle of excluded middle allows us to declare, nevertheless, that it landed either heads or tails. It also allows us to define a number that equals 2 if the coin landed tails and 4 if it landed heads and then, for instance, to prove that this number is even. Of course, we have no way of ever knowing the value of the number; for this, we would have to find the coin in the sand at the bottom of the ocean. So it is the excluded middle that allows us to build proofs of existence that do not produce witnesses. This is the reason why proofs that do not rely on the excluded middle are called "constructive proofs."

BROUWER'S CONSTRUCTIVISM

It is hard to date the first genuine nonconstructive demonstration in history, because some mathematicians may have used the principle of excluded middle out of convenience, even when it would be easy to reformulate their arguments as constructive proofs. It is generally agreed that the first proofs that are truly nonconstructive date back to the late nineteenth century. At that time, mathematics took a leap into abstraction, especially with the development of set theory and the beginnings of topology. The emergence of nonconstructive proofs caused an uneasy stir in the mathematical community, and some mathematicians, such as Leopold Kronecker and Henri Poincaré, expressed their doubts about this modern mathematics: they were suspicious of its objects, which they found too abstract, and the new methods it used, which they judged dubious on account of their ability to prove theorems of existence without providing a witness. It was in this atmosphere of distrust that, in the early twentieth century, a radical program was formulated by Luitzen Egbertus Jan Brouwer: his plan was to do away with the principle of excluded middle in mathematical proofs – which meant rejecting the proof of the existence of a last station before the border. This "constructivist" (or "intuitionist") program put mathematicians in an awkward position: some propositions, whose known proofs relied on the excluded middle, were then regarded as theorems by some but not by others.

Such a crisis is as embarrassing as it is unusual – it is not, however, unique. The Greeks refused to add up infinite sequences of numbers, whereas sixteenth century mathematicians were quite willing to do it. Besides, as we are about to see, another crisis, called "the crisis of non-Euclidean geometries," took place at the beginning of the nineteenth century, and yet another was to occur in the late twentieth century when computer-aided proofs appeared. All the same, the situation was problematic and the crisis needed to be resolved.

Constructivists did not all have the same objection to the principle of excluded middle. The moderates – among whom is our explorer – thought a nonconstructivist proof was of little interest: what is the point of knowing there exists a last station before the border if we don't know which station it is? The moderates rejected nonconstructive proofs not on grounds of incorrectness but of uselessness. A yet more moderate view suggests that, because constructive proofs yield more information than nonconstructive proofs, whenever possible we should favor constructive demonstrations. Conversely, a much more radical view holds that nonconstructive proofs are false. This point of view is Brouwer's – it is only a minor overstatement to say that he would have thought twice about crossing a bridge if its engineers had used the excluded middle to prove that it could bear his weight.

Several things fanned the flames of the crisis. First, by attacking nonconstructive methods, Brouwer disparaged one of the greatest mathematicians of the early twentieth century, author of several nonconstructive proofs, namely Hilbert. The mathematical disagreement escalated into a personal conflict between Brouwer and Hilbert, so that neither really tried to listen to the arguments of the other. Second, besides this quarrel about the principle of excluded middle, other disputes broke out. For instance, Brouwer upheld the idea that our intuition about mathematical objects matters more than the knowledge we gain about those objects through proofs – hence the name "intuitionism." This position might have tipped the scales in favor of a moderate thesis, according to which intuition plays a key part in the construction of new knowledge, and proof only serves to validate this intuition. However, Brouwer's intuitionist mysticism rendered his ideas unpopular – constructivists were soon perceived as members of an esoteric cult. Even today, we can find traces of such opinions: in a text published in 1982 by Jean Dieudonné, constructivists are compared to "some backward religious sect still believing the earth to be flat."

RESOLVING THE CRISIS

The crisis of constructivism was eventually resolved. Nowadays, there is no longer cause to wonder whether it is acceptable to use the excluded middle in mathematical reasoning: the question must be rephrased.

The better to understand how the crisis was resolved, let's compare it to another crisis that hit the mathematical community a century earlier: the crisis of non-Euclidean geometries.

Among the axioms of geometry is one that has been the cause of much controversy ever since antiquity: "the axiom of parallels." In its modernized form, this axiom states that, given a line and a point outside it, there is exactly one line through the given point that is parallel to the given line. For many geometers, this proposition was not obvious enough to be accepted as an axiom: it was seen, on the contrary, as a proposition that needed to be proved from the other axioms of geometry. The fact that Euclid failed to do so is not reason enough to accept it as an axiom.

From Euclid's day to the early nineteenth century, many mathematicians tried to prove this proposition from the other axioms of geometry, to no avail. Many of these attempts resorted to reductio ad absurdum: they posed the negation of the proposition as a hypothetical axiom and tried to reach a contradiction. Mathematicians elaborated theories in which they formulated the hypothesis that, given a line and a point outside it, there were several lines through the given point lying parallel to the given line, or none at all. The theories thus reached did not seem contradictory; in fact, they seemed quite interesting.

So, by the beginning of the nineteenth century, Carl Friedrich Gauss and later Nicola Lobatchevsky, János Bolyai, Bernhard Riemann, and others offered alternative geometries based on axioms other than the axiom of parallels. For instance, in Riemann's geometry, given a line and a point outside it, there is no line through the given point lying parallel to the given line. Thus, in the early nineteenth century, mathematicians began to question whether this or that axiom could really be used in geometry – just as they would question, a century later, whether this or that rule of inference could be used in proofs. This issue was so controversial that Gauss even decided not to publicize his work on the subject so as not to scare his contemporaries.

Resolving the crisis of non-Euclidean geometries was Poincaré's aim when he suggested that the axioms of geometry were not statements of the obvious but rather implicit definitions of "point" and "line," and that different mathematicians use different axioms because they assign different meanings to these words. The alternative notions of "point" and "line" defined by non-Euclidean geometries soon turned out to be much less exotic than they seemed. For example, whereas, in Euclidean geometry, the sum of the angles in a triangle is always 180°, in Riemann's geometry, this sum is always bigger than 180°. On the surface of the earth, the triangle whose summit is situated in the North Pole and whose base runs along the equator from a longitude of 0° to 90° has three right angles, the sum of which is 270°. The lines drawn across a curved surface, like that of the Earth, are an instance of what Riemann's geometry defines as "straight lines."

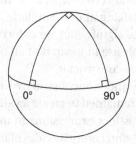

Like axioms, inference rules are not obvious facts that stand to reason. They are implicit definitions of the conjunctions and quantifiers that appear in those rules: "and," "or," "if…then," "there exists," and so on. So the fact that the constructivists did not use the same inference rules as other mathematicians has a simple explanation: they assigned different meanings to the words "and," "or," "if…then," and "there exists." In particular, the proposition "there exists an object that…" meant "we know of an object that…" to the constructivists and "there must exist an object that…even if we don't know the object" to other mathematicians. We now understand why the proposition "there exists a last station before the border" is false for those who take it to mean "we know the last station before the border" and true for those who believe it means "there must exist a last station before the border even if we don't know which it is."

With hindsight, the crisis of constructivism had the merit of drawing attention to the different shades of meaning of the common phrase "there exists." Actually, it is not so unusual for an imprecise everyday notion to give birth to several precise, mathematical notions – thus did the notion of "number" generate the notions of whole number, real number, and so forth.

CONSTRUCTIVITY TODAY

The two variants of the phrase "there exists" are not necessarily mutually exclusive. We can use the word "number" to designate a whole number one day and to refer to a real number the next. All we need to do is add the adjective "whole" or "real" to specify what kind of number we are talking about. Similarly, we can use the phrase "there exists" now to signify that we can construct an object, and later use the same phrase to mean that an object must exist even if we don't know how to construct it.

For constructive and nonconstructive mathematics to finally make peace, it only remained to create a variation of predicate logic that would assign distinct expressions to the two meanings of the phrase "there exists" and comprise rules of inference clarifying their meanings. One such logic was put forward by Gödel in 1933, under the name "negative translation." The details of this logic matter less than the fact that it proved constructive and nonconstructive mathematics could coexist peacefully, contrary to what Hilbert and Brouwer thought.

After the crisis was resolved, some mathematicians remained uninterested in the notion of constructivity. They kept on using the principle of excluded middle – in other words, they kept on using only one of the two meanings of the phrase "there exists." Others kept on rejecting the excluded middle and used only the other meaning of the phrase "there exists." Others yet adopted a more open attitude and used both meanings of the phrase, using the principle of excluded middle when necessary – after all, once it had allowed them to prove their theorem, nothing prevented them from proving the same theorem again, this time in a constructive manner.

CHAPTER EIGHT

Constructive Proofs and Algorithms

WHAT DOES THE NOTION of constructive proof have to do with the subject of this book, namely computation? The notions of computation and algorithm did not originally play such an important part in the theory of constructivity as in, say, the theory of computability. However, behind the notion of constructive proof lay the notion of algorithm.

CUT ELIMINATION

We have seen that proofs of existence resting on the principle of excluded middle do not always contain a witness. On the other hand, a proof of existence that does not use the excluded middle always seems to contain one, either explicitly or implicitly. Is it always so, and can this be demonstrated?

Naturally, the possibility of finding a witness in a proof does not depend solely on whether the demonstration uses the excluded middle; it also depends on the axioms used. For example, an axiom of the form "there exists..." is a proof of existence in itself, yet this proof produces no witness. The question of the presence or absence of witnesses in proofs that do not use the excluded middle thus branches into as many questions as there are theories – arithmetic, geometry, Church's type theory, set theory, and, to begin with, the simplest of all theories: the axiom-free theory.

One of the first demonstrations that a proof of existence constructed without calling upon the principle of excluded middle always includes a witness (at least implicitly) rests on an algorithm put forward in 1935 by Gerhard Gentzen, called the "cut elimination algorithm." This algorithm, unlike those we have mentioned up to this point, is not applied to numbers, functional expressions, or computation rules – it is applied to proofs.

A proof can contain convoluted arguments, which are called "cuts." Gentzen's algorithm aims at eliminating those cuts by reorganizing the proof in a more straightforward way. For example, when a proof proceeds by demonstrating a general, intermediate result, only to apply it to a specific case, Gentzen's algorithm replaces that detour in the demonstration with a direct proof of the specific case. When a proof of existence that does not rely on the excluded middle is thus reorganized, the proof produced always ends with an introduction rule of the existential quantifier and, in that case, the witness is explicit in the demonstration.

The link between the notion of proof – constructive or not – and that of algorithm becomes apparent, because proofs, like numbers, are objects we can compute with, especially when we apply to them this cut elimination algorithm.

Gentzen's algorithm was applied to proofs in simple theories, such as the axiom-free theory and arithmetic. In 1970, Jean-Yves Girard extended it to Church's type theory, a variation on set theory. In years that followed, it was further expanded and came to be applied to more and more elaborate theories.

FUNCTIONS AND ALGORITHMS

We have seen that functions can sometimes be defined by functional expressions, such as $x \mapsto x \times x$, but this is not always the case. Unlike the function that, to a number, assigns the square of that number, the function that, to a number, assigns the square root of that number cannot be defined by such an expression. To define this function, we must first define the relation between x and its image y, that is $x = y \times y$, and then prove that for each positive real number x there exists one and only one positive real number y related to x in this way. According to whether the principle of excluded middle is an option, the number of functions we can define varies. The theory of constructivity thus allows us to classify functions into two categories: those that can be defined by constructive proofs and those whose definitions require nonconstructive proofs.

What is the advantage of defining a function in a constructive way? The payoff is an algorithm that, when applied to an element a of the domain of definition of the function f, computes the object b assigned to a by this function. This theorem, which we owe to Kleene, can be seen as a simple consequence of the cut elimination theorem. Indeed, since for all elements x of the domain A, there exists a unique object y assigned to x by the function, we can infer that there exists, in particular, an object assigned to a by that function; Gentzen's algorithm, when applied to this existence proof, produces a witness of that existence, namely the object b assigned to a by the function.

Hence, the notion of algorithm receives a new, more abstract definition: an algorithm is a function that can be defined in a constructive way.

CONSTRUCTIVE PROOFS AS ALGORITHMS

Today, these two links between the notions of constructive proof and algorithm are recognized as the tip of an vast iceberg: the "algorithmic interpretation" of constructive proofs. In the twenties, the first green shoots of this idea could already be found in the writings of Luitzen Egbertus Jan Brouwer, Arend Heyting, and Andre Kolmogorov, but its main development took place in the sixties with the work of Haskell Curry, Nicolaas Govert de Bruijn, and William Howard.

In mathematics, the words "interpretation" and "definition" differ only slightly in meaning, but they are not synonyms: a definition defines a new notion whereas an interpretation redefines a notion that already exists as a prime notion, or as a notion with an unsatisfactory definition. For example, the notion of complex number has existed since the late Middle Ages as a prime notion, but it was not until the nineteenth century that complex numbers were interpreted as pairs of real numbers, points in the plane, and so on; in other words, only then did they receive a proper definition. Interpreting the notion of proof therefore means providing a new answer to the question: what is a proof?

To respond to this question, we must first answer two related questions: how are proofs constructed? And how are they used?

When we have a proposition of the form "if A then B," its proof can be used to infer proposition B from proposition A – in other words, to construct a proof of B with a proof of A or to transform a proof of A into a proof of B. A proof of the proposition "if A then B" is used in the same way as an algorithm that transforms proofs of proposition A into proofs of proposition B. Moreover, it can be shown that a proof of the proposition "if A then B" is constructed in the same way as the algorithm. We might as well call these proofs algorithms and define them as such.

This algorithmic interpretation of proofs reverses the perspective of the axiomatic method that, since Euclid, has founded mathematics on the notion of proof. All of a sudden, it turns out that this notion of proof is not a prime notion but can be defined in terms of a more fundamental notion: that of algorithm. By basing their whole mathematics on the notion of algorithm, the Mesopotamians had – unbeknownst to them – started with the most fundamental notion. It was the Greeks who were ill-advised to found their mathematics on the notion of proof.

In the algorithmic interpretation of proofs, a proof of the proposition "for all numbers x and y, there exists a number z that is the greatest common divisor of x and y" is an algorithm that, to all pairs of numbers a and b, assigns a pair composed of a number c and a proof that c is the greatest common divisor of a and b. When this algorithm is applied to the pair formed of 90 and 21, after carrying out a computing operation, we obtain the number 3 as well as a proof that 3 is indeed the greatest common divisor of 90 and 21.

What is the computing operation that we must carry out to obtain that result? First, we must execute the computation task that Gentzen's cut elimination algorithm would have carried out in order to simplify the proof. Algorithmic interpretation of proofs thus shows not only that proofs are algorithms, but also that cut elimination acts as an interpreter for these algorithms.

Besides, the computing operation is close to the one Euclid's algorithm would have led to. In Chapter 2, we suggested that the apparent contradiction between mathematical discourse – which, since the Greeks, tends to ignore computation – and mathematical practice – which greatly depends on it – can be explained away by the fact that Euclid's algorithm, for instance, when applied to two numbers, not only enables us to compute their greatest common divisor but also produces a proof that this number is indeed their greatest common divisor. Algorithmic interpretation sheds light on this old link between proofs and algorithms: a constructive proof of the existence of a greatest common divisor is an algorithm that both computes a number and proves this number to be the greatest common divisor of the pair the algorithm is applied to.

To conclude, the link between the notions of constructive proof and algorithm turns out to be rather simple after all: constructive proofs are algorithms.

These two chapters dedicated to the theory of constructivity end, again, with a paradox. Like the theory of computability, constructivity attaches great importance to the notions of algorithm and computation, because it eventually defines constructive proofs as algorithms. But the proofs with which the theory is concerned are proofs in predicate logic, proofs built with axioms and rules of inference in keeping with the axiomatic conception of mathematics, proofs that allow no room for computation.

From the dawn of the twentieth century to the late sixties, neither the theory of computability nor that of constructivism challenged the axiomatic method. Until the late sixties, proofs still required axioms and rules of inference and still made no room for computation. However, because of the role they ascribe to the notion of algorithm, the two theories paved the way for the critique of the axiomatic method, which began in the following decade.

PART THREE

Crisis of the Axiomatic Method

CHAPTER NINE

Intuitionistic Type Theory

IT WOULD NOT BE UNTIL THE EARLY 1970S that the axiomatic method would be challenged. Then, surprisingly enough, it was called into question simultaneously and independently in several branches of mathematics and computer science. Few of the main players in this episode in the history of mathematics were aware that they were pursuing the same aim, within logic as researchers pushed forward the work of their predecessors on constructivity, within computer science, and within "real-world" mathematics. This chapter will focus on logic.

INTUITIONISTIC TYPE THEORY

In the late sixties, many breakthroughs sparked a revival of interest in constructivity. On one hand, the algorithmic interpretation of proofs was developed thanks to the work of Curry, de Bruijn, and Howard; on the other hand, William Tait, Per Martin-Löf, and Jean-Yves Girard proved cut elimination for new theories. Most important, Girard proved cut elimination for Church's type theory, a variation on set theory. It thus became possible to provide constructive mathematics with a general framework equivalent to Church's type theory or set theory. Martin-Löf offered one such framework: intuitionistic type theory.

Intuitionistic type theory was born of an ascetic approach to logic: in order to provide a minimal basis for mathematics, this theory aims not only to exclude the principle of excluded middle, but also to break free of three axioms in Church's type theory, which we will not explore in detail in this book, namely the axiom of extensionality, the axiom of choice, and the axiom of impredicative comprehension. In the early seventies, many mathematicians doubted, understandably, that a theory so weakened would be capable of expressing much at all. Thirty years later, however, we are forced to recognize that vast sections of mathematics have been successfully expressed in this theory and in some of its extensions, such as Thierry Coquand and Gérard Huet's calculus of constructions, for instance.

EQUALITY BY DEFINITION

Martin-Löf's intuitionistic type theory is more than just a variation on Church's type theory, or on set theory, give or take a few axioms. It also introduces new ideas and concepts. For example, it integrates the notion of algorithmic interpretation of proof by defining proofs as algorithms, and it introduces a notion absent from both Church's type theory and from set theory, namely that of "equality by definition."

Church's type theory, like set theory, recognizes only one notion of equality: we can convey the fact that two expressions refer to the same object; but whether this fact follows simply from a definition or whether it requires a complex reasoning process, the way it is conveyed is the same. Intuitionistic type theory, on the other hand, has two notions of equality: the ordinary notion of equality present in set theory and in Church's type theory, and this new notion of equality by definition.

The simplest conceivable definition mechanism consists of adding a symbol to a language and declaring it equal to a certain expression. For example, instead of repeatedly copying the expression $1/2$, we can decide to give it a shorter name, say h. In other words, we introduce a symbol, h, and declare it equal to $1/2$. Then, the expression $h + 1$ and the expression $1/2 + 1$ (where the symbol h is merely replaced with the expression it stands for, $1/2$) are said to be equal "by definition." Similarly, the propositions "$h + 1 = 3/2$" and "$1/2 + 1 = 3/2$" are equal by definition.

In order to add such a definition mechanism to a theory, it is necessary to adapt the notion of proof. This can be done in two ways. The first involves adding an axiom $h = 1/2$ or inference rules enabling us to deduce the proposition "$h + 1 = 3/2$" from the proposition "$1/2 + 1 = 3/2$," and vice versa. Once this axiom or these rules have been added, if one of these propositions is provable, then so is the other; their proofs, however, differ: proving one requires proving the other and adding a deduction step to the demonstration. The second possibility consists in declaring that proving one of these propositions amounts to proving the other – this is how things work in intuitionistic type theory. Thus, definitions are neither axioms nor inference rules, but a third ingredient enabling the construction of proofs.

Actually, there is more to the definition mechanism of intuitionistic type theory than just replacing a symbol with an expression. We mentioned the beta-reduction rule, which allows us, for instance, to transform the expression $(x \mapsto (x \times x))(4)$ into 4×4; recall the conundrum it posed for Church, whose first instinct was to regard this transformation as a computation stage but who later abandoned this idea, retreated to a more traditional approach, and laid out an axiom simply stating the equality of the two expressions. With intuitionistic type theory, Martin-Löf suggested that these two expressions are equal because of the definition of the symbol $\mapsto$. This definition is more complex than the one that replaces the symbol h with $1/2$; however, the definition of the symbol $\mapsto$ does seem to lie in the fact that the expression $(x \mapsto t)(u)$ is by definition equal to the expression t where the variable x is replaced with the expression u. Thus, the propositions $(x \mapsto (x \times x))(4) = 16$ and $4 \times 4 = 16$ are equal by definition, and proving one amounts to proving the other. Intuitionistic type theory goes even further and brands "equal by definition" other mechanisms, such as definition by induction, so that the expressions "$2 + 2$" and "4" are considered equal due to the definition of addition. Two apparently distinct notions thus surprisingly meet, namely that of "equality by definition" and that of "equality by computation." Actually, in intuitionistic type theory, "equality by definition" might as well be called "equality by computation."

EQUALITY BY DEFINITION AND ANALYTIC JUDGMENTS

This notion of "equality by definition" does not go quite as far, however, as the notion that underlies Poincaré's idea that the axioms of geometry are implicit definitions of the notions of point and straight line. In his notion of definition, whenever two things can be proved equal, they are so by definition, because they are equal as a consequence of the axioms. So that, in the end, there appears to be little difference between equality by definition and regular equality. Besides, in intuitionistic type theory, equality by definition is decidable, whereas Church's theorem shows that equality by implicit definition is not.

A major event occurred between the findings of Poincaré and those of Martin-Löf: Church's theorem was developed along with the theory of computability. Church's theorem makes it necessary to rethink the notion of definition. The common conception seems to assume that equality by definition is decidable, which leaves us no choice but to discard Poincaré's idea according to which axioms are implicit definitions. More precisely, it forces us to distinguish between two notions of definition: the common concept of definition, which assumes that equality by definition is decidable, and a more liberal concept, which implies no such thing and is not to be confused with the first. Whether axioms are implicit definitions or not is actually merely a matter of terminology.

By transforming the theory of definitions in this way, the development of the theory of computability also transforms our perception of other notions, namely those of analytic and synthetic judgment, which were used by Kant and Frege at a time when the theory of computability did not yet exist. In intuitionistic type theory, a judgment is analytic when it requires only computation, and it is synthetic when it demands a demonstration. Thus, the judgment according to which $2 + 2$ equals 4 is analytic, but judging the proposition "the sum of the angles in a triangle is $180°$" to be true is synthetic, even though this proposition is true by necessity and does not express anything regarding nature. That judgment is synthetic, just like the judgment that "the earth has a satellite" is true, because judging either one of these propositions true requires more than simple computation. In the first case, we must construct a proof, and in the second, we need to make observations. Using this notion of analytic judgment brings us to approximately the same conclusion as Kant reached: mathematical judgments are, except in a few cases, synthetic.

In Chapter 3, we reached another, different conclusion. It may seem strange that mathematical judgments can be deemed now synthetic, now analytic. This is not, however, a contradiction: it is simply a difference in terminology. The classical terminology, which opposes analytic judgments to synthetic judgments, is misleading. It gives the impression that there exist two types of judgments. Actually, there exist at least three: judgments based on computation, those based on demonstration, and those requiring some interaction with nature or observation. There is a general agreement that mathematical judgments belong to the second category – it is merely the name of that category that is debated.

SHORTER PROOFS, LONGER PROOF CHECKING

In intuitionistic type theory, proofs of the statement "$2 + 2 = 4$" are the same as proofs of the statement "$4 = 4$," because the two propositions are equal by definition. The proof of a proposition like "$2 + 2 = 4$" is therefore very short: it rests on the axiom according to which "for all x, $x = x$," which is applied to the number 4. This proof is short to write because it has been stripped of all the computation stages between $2 + 2$ and 4. On the other hand, upon reading this demonstration, if we wish to ascertain that it is, indeed, a proof of the proposition "$2 + 2 = 4$," we need to carry out again all the unwritten computations.

Depending on how you define a notion, proving a proposition may take more or less time to write and check. Take the example of composite numbers. A number is said to be composite when it is not prime, that is, when it can be divided by a number other than 1 or itself. For example, the number 91 is composite because it can be divided by the number 7. An algorithm exists that determines whether a number is composite: try to divide the given number by all natural numbers smaller than it. We can therefore define an algorithm f that produces the value 1 if it is applied to a composite number, and 0 otherwise. The fact that 91 is a composite number is then expressed by the proposition "$f(91) = 1$." This proposition is equal by definition to the proposition "$1 = 1$" and its proof is very short to write: we need only cite the axiom "for all x, $x = x$." Verifying that this proof is a demonstration of the proposition $f(91) = 1$, however, requires computing $f(91)$ all over again, that is, retesting the divisibility of 91 by all smaller natural numbers.

For another solution we can define an algorithm g that applies to two numbers x and y and produces the value 1 when y is a divisor of x, and 0 when it is not. The fact that 91 is composite is then expressed by the proposition "there exists y such that $g(91, y) = 1$." The proof of this proposition is a little longer: it rests on the introduction rule for existential quantification, which it applies to a divisor of 91, say 7; then it demonstrates the proposition "$g(91, y) = 1$," which is equal by definition to the proposition "$1 = 1$" and which is proved as demonstrated earlier. On the other hand, this proof is quick to check, because there is only one operation that needs to be performed again: we must compute $g(91, 7)$, and to do so, all that is required is to test the divisibility of 91 by 7. So, as far as the length of the demonstration and the time needed to verify it are concerned, we seem to have found a happy medium with this proof.

There are other satisfactory compromises. We could, for instance, define an algorithm h that applies to three numbers x, y, and z, and gives the value 1 if x is the product of y and z, and 0 otherwise. The fact that 91 is composite is then expressed by the proposition "there exist y and z such that $h(91, y, z) = 1$." Proving this proposition takes longer. First, the existential quantification introduction rule is applied to two numbers whose product is 91, say 7 and 13. This produces the proposition "$h(91, 7, 13) = 1$," which is equal by definition to the proposition "$1 = 1$" and is demonstrated as earlier. This lengthy demonstration is balanced, however, by its very speedy verification: we simply recompute $h(91, 7, 13) = z$ – in other words, multiply 7 by 13, compare the result to 91, and Bob's your uncle!

Last but not least, let's not forget the proof that does without computation rules altogether. This proof takes even longer to construct, because it decomposes the multiplication algorithm, applied to 7 and 13, into its most minute stages. Although this proof may be easy to check, it is so long that checking it, too, turns into a tedious, drawn-out process.

To illustrate the differences between these four proofs, imagine a mathematician, desirous of knowing whether 91 is a composite number but too busy to solve the problem herself. She asks a colleague to do it for her. The colleague can provide her with four distinct answers:

- 91 is a composite number, as you will find by doing the required operations yourself;
- 91 is a composite number, because it is divisible by 7;
- 91 is a composite number, because it is the product of 7 and 13;
- 91 is a composite number, because it is the product of 7 and 13; indeed, three times seven is twenty-one, I put down one and carry two, seven times one is seven, plus two is nine, put down nine: ninety-one.

Obviously, the best answers are the second and third – the first being too laconic and the last, too verbose.

It is interesting to note that, whereas the only answer available to prehistoric mathematicians is the first, in the axiomatic method, the only possible answer is the last. Answers two and three are only possible because, in intuitionistic type theory, proofs can be constructed with axioms, inference rules, and computation rules.

Thus, in the early 1970s, the notion of computation was introduced in Martin-Löf's intuitionistic type theory through the notion of "equality by definition" – yet mathematicians did not take in the full extent of this revolutionary event. In this theory, besides rules of inference and axioms, they could finally use a third ingredient in building proofs, namely computation rules.

CHAPTER TEN

Automated Theorem Proving

IN THE EARLY SEVENTIES, when word of Martin-Löf's type theory and subsequent research on the subject had not yet reached computer scientists, the idea that a proof is not built solely with axioms and inference rules but also with computation rules developed in computer science – especially in a branch called "automated theorem proving." Thus, work on type theory and work on automated theorem proving were pursued contemporaneously by two distinct schools that ignored each other – admittedly, an improvement on the behavior of the schools that developed, respectively, the theory of computability and the theory of constructivity, which abused each other. It was not until much later that the links between these projects were finally revealed.

An automated theorem proving program is a computer program that, given a collection of axioms and a proposition, searches for a proof of this proposition using these axioms.

Of course, Church's theorem sets a limit on this project from the outset, because it is impossible for a program to decide whether the proposition that it has been requested to supply with a proof actually has one or not. However, there is nothing to prevent a program from searching for a proof, halting if it finds one, and continuing to look as long as it does not.

THE FANTASY OF "INTELLIGENT MACHINES"

In 1957, during one of the first conferences held about automated theorem proving, the pioneers in the field made some extravagant claims. First, within ten years, computers would be better than

humans at constructing proofs and, as a consequence, mathematicians would all be thrown out of work. Second, endowed with the
ability to build demonstrations, computers would become "intelligent": they would soon surpass humans at chess, poetry, foreign
languages, you name it. Cheap sci-fi novels began to depict a world
in which computers far more intelligent than humans ruthlessly
enslaved them. Ten years went by and everyone had to face the
facts: neither of these prophecies were even close to being fulfilled. The hyperbole surrounding the project of designing intelligent machines, and the fears and disappointments thus engen-
dered, had the effect of shrouding in confusion the more measured
goal of automated theorem proving and the questions it entails.

One of these questions concerns the theoretical possibility that
a machine can build proofs as well as a human being. If we accept
the psychological form of Church's thesis, it seems to follow that the
mental processes at work when constructing a proof can, in theory, be simulated by a set of computation rules – here by a computer. Because the psychological form of Church's thesis is only a
hypothesis, we could instead adopt the opposite thesis, according
to which an uncrossable boundary separates human beings from
machines and human beings will always be better at building proofs
than computers. However, we have seen that the psychological form
of Church's thesis is a consequence of two other theses, namely the
physical form of Church's thesis and the materialist thesis, which
states that human beings are part of nature. If we want to contradict the psychological Church thesis, we must abandon at least one
of these other two theses, either the one stating human beings are a
part of nature, which would lead to the conclusion that the mind is
something distinct from the functioning of the brain, or the physical form of Church's thesis, in which case one would give up either
the idea that the density of information is finite, or that its transmission speed is. Some like Roger Penrose, hope the creation of a
new physics that challenges principles such as the finite density of
information and its finite transmission speed and explains why the
functioning of the brain cannot be simulated by a computer.

These leads are worthy of attention. However it seems impossible to claim at once that information has finite density and transmission speed, that human beings are part of nature, and that the

mental processes at work in the building of a proof cannot, in theory, be simulated by a computer.

Automated theorem proving raises a second question, relatively independent of the first: can machines build demonstrations as well as human beings, in the current state of things? This question is more easily answered. Even if we believe it to be possible, in theory, for a machine to build proofs as well as a human being, we're bound to acknowledge that the performance of automated proof programs are as yet inferior to that of human beings.

The last question concerns our fears about efficient proof-building machines: are those fears justified or not? First, even if those fears were justified, it would leave the answer to the two previous questions unchanged. Defending the thesis that machines cannot build proofs as well as human beings because it would be disagreeable is tantamount to claiming hemlock cannot be a poison because it would be unpleasant. As for determining whether these fears are justified or not, it is hard to tell; although machines are already better than us at doing multiplications, playing chess, and lifting heavy loads, no calculator, chess program, or crane has seized power so far. So, is there reason to fear a machine that constructs demonstrations better than we do?

For a long time the fascination exerted by the idea of a competition opposing human beings and machines, as well as the desire to settle decisively whether it is possible for machines to reason like human beings, has concealed a much more interesting fact: automated proof programs have made steady and significant progress since the 1950s. There are myriad proof-building processes, which construct demonstrations of varying complexity, and each new generation of program has proved propositions that had been stumbling blocks for the previous generation. It is much more interesting to try to understand the ideas that made such progress possible than to figure out whether a machine can reason better than a human being.

"RESOLUTION" AND PARAMODULATION

In its early stages, automated proof inherited conceptual frameworks from logic and, more specifically, from the axiomatic method and predicate logic. Thus, the first methods of automated proof, such as "resolution," devised by Alan Robinson in 1965, and paramodulation, developed by Larry Wos and George Robinson in 1969, were used to search for proofs within predicate logic. At the core of these methods lies an algorithm, namely the unification algorithm, that compares two expressions and suggests expressions with which the variables can be replaced so as to make the two expressions identical. For instance, on comparing the expressions $x + (y + z)$ and $a + ((b + c) + d)$, the unification algorithm suggests substituting the expression a for the variable x, the expression $b + c$ for the variable y, and the expression d for the variable z, making the two expressions identical. By contrast, the algorithm must fail when comparing the expressions $x + (y + z)$ and a, because there is no way of making these two expressions identical by replacing the variables x, y, and z.

When we try to prove the proposition $a + ((b + c) + d) = ((a + b) + c) + d$ using the axiom "for all x, y, and z, $x + (y + z) = (x + y) + z$," which states that addition is an associative operation, paramodulation suggests comparing the expression $x + (y + z)$, that is, one side of the equality stated by the axiom, with all the expressions in the proposition we wish to demonstrate. When this comparison is successful and finds expressions that can be substituted for the variables, we can perform the same substitutions in the other side of the axiom. In our example, to compare ("unify") $x + (y + z)$ with $a + ((b + c) + d)$, we can substitute a for x, $b + c$ for y, and d for z. The axiom then translates to $a + ((b + c) + d) = (a + (b + c)) + d$, and the proposition to be proved becomes $(a + (b + c)) + d = ((a + b) + c) + d$. In the second step, we replace $a + (b + c)$ with $(a + b) + c$, and the proposition now becomes $((a + b) + c) + d = ((a + b) + c) + d$, which can easily be demonstrated, because it is of the form "$x = x$."

The main idea behind resolution and paramodulation is using the unification algorithm to suggest expressions with which the variables can be replaced. Before these methods were devised, the approach was to blindly replace the variable by every possible expression, hoping the right substitution would eventually be found – and indeed, it eventually was, but this often took quite a while. With hindsight, it appears that the idea of unification had already been put forward by Herbrand in 1931, long before Robinson's day: he had conceived it while doing research on Hilbert's decision problem.

Unification problems are similar in many ways to equations. In both cases, we proceed by giving values to variables for the purpose of making two things equal. The specificity of unification lies in the form of the equality it seeks to achieve. When unifying two expressions such as $x + 2$ and $2 + 2$, we need to replace the variable x so as to make the two expressions identical. Here, the solution is to replace the variable x by 2. In other cases, however, the unification problem cannot be solved. There is no way to unify the two expressions $x + 2$ and 4, for example: if you replace x by 2, you get the expression $2 + 2$, which is not identical to the expression 4; the fact that $2 + 2$ equals 4 is irrelevant.

TURNING AXIOMS OF EQUALITY INTO COMPUTATION RULES

Given an axiom of the form $t = u$, paramodulation allows us to replace any instance of t with the corresponding instance of u, and vice versa. Thus, using the associativity axiom, in expressions of the form $p + (q + r)$, we can shift the parentheses to the left and, in expressions of the form $(p + q) + r$, we can shift them to the right.

This method often demands long computation processes, even to solve simple problems. Continuing to use the associativity axiom as an example, if we want to prove a more complex proposition than the previous one, say, the proposition $((a + (b + c)) + ((d + e) + (f + g)) + h) = ((a + b) + (c + d) + (e + ((f + g) + h)))$, there are more than a dozen ways to approach the problem. Every time there is an addition in the proposition we aim to prove and one of the terms of this addition is itself an addition, the parentheses may be shifted. Exploring each and every possibility for this step and for all the following steps takes a lot of time, even for a computer: this limits the method's chances of success. Indeed, if solving even the simple problem we are considering here lasts several minutes, the method is pointless.

Yet there exists a simpler way of solving this problem: always shifting the parentheses to the left, never to the right. By following this strategy, the proposition we wish to prove is transformed into

$(((((((a + b) + c) + d) + e) + f) + g) + h = ((((((a + b) + c) + d) + e) + f) + g) + h$, which is easy to demonstrate because it is of the form $x = x$. More generally, every time we have an axiom of the form $t = u$, we can decide to apply it only in one direction – for instance, we may choose to replace t with u, but not u with t. This restriction eliminates all cycles in which u is substituted for t, then t for u, which would bring us back to the starting point. When we decide to use an axiom of the form $t = u$ but restrict ourselves to replacing t with u but never u with t, we turn that axiom into a computation rule.

Unfortunately, this method does not suffice for, in propositions such as the one quoted earlier, there exist numerous different ways of shifting the parentheses to the left. To avoid exploring them all, one must use the fact that the result of the computation does not depend on the order in which one shifts these parentheses.

It is natural to wish for the result of a calculation to be independent of the order in which the computation rules are applied. This may or may not be the case, as you try different rules. For instance, if two computation rules respectively allow you to transform $0 + x$ into x and $x + x$ into $2 \times x$, and assuming you start with $0 + 0$, then you reach 0 if you apply the first rule and 2×0 if you apply the second one. When the final result is independent of the order of the operations, the set made up of the computation rules used in the process is called "confluent." This property is also called the Church and Rosser property, for it was one of the very first properties of beta-reduction which they established in the thirties. When a set of computation rules like this one is not confluent, it is sometimes possible to add rules to it so as to make it confluent. In the present case, all one needs to do for the system to become confluent is add a third rule turning 2×0 into 0.

It was Donald Knuth and Peter Bendix who, in 1970, suggested transforming axioms of the form $t = u$ not only into computation rules but also into rules forming a confluent set. This enables one to conceive automated proof methods that are quicker than the previous models at demonstrating propositions such as the earlier one. Indeed, unlike their prototypes, the new methods do not take detours, shifting parentheses now to the left, now to the right, nor do they multiply redundant attempts at solving the problem which only differ from each other in the rules' application order.

FROM UNIFICATION TO EQUATION SOLVING

The problem when one transforms axioms into computation rules is that one must renounce the possibility of proving certain propositions formed with quantifiers – for example, the proposition "there exists y such that $a + y = (a + b) + c$" was provable with the previous methods, which still laid out associativity as an axiom. Indeed, one cannot apply the associativity computation rule to this proposition, for all parentheses are already as far left as can be, and the unification of the expressions $a + y$ and $(a + b) + c$ cannot but fail for, no matter what expression y is replaced with in $a + y$, one can never reach the result $(a + b) + c$.

In 1972, Gordon Plotkin laid the groundwork of a method allowing one to turn axioms into computation rules without leaving out any type of proof. In order to prove the aforementioned proposition, Plotkin proceeded by comparing the expressions $a + y$ and $(a + b) + c$, as was done in the previous methods. The difference is that, whereas those methods failed, Plotkin's produces the solution $b + c$: indeed, if one replaces the variable y by the expression $b + c$ in the expression $a + y$, one obtains $a + (b + c)$, which is not, of course, identical to $(a + b) + c$, but which can be turned into $(a + b) + c$ thanks to the rule of associativity. Plotkin's unification algorithm is clearly more complex than Robinson's, since it takes computation rules into account – to put it in Plotkin's own words, the associativity axiom is "built in" to the unification algorithm. Further advances have shown that other axioms of equality besides the associativity axiom could be built in to the unification algorithm.

For instance, the unification algorithm can build in all arithmetic computation rules; thus the unification problem $x + 2 = 4$, which could previously find no solution without resorting to computation rules, is now provided with one, since $2 + 2$ can be computed into 4. These extended unification problems reveal the existence of another tool at one's disposal to solve mathematical problems besides reasoning and computation, namely equation solving. Thus, as all high school students have experienced, to solve a problem, sometimes, one needs to carry out a computation operation, sometimes, one must build a reasoning process, and sometimes it is necessary to resolve an equation.

This irruption of equation solving in the realm of automated proof methods leads us to look with new eyes at the equations used in mathematics. At first glance, equations are formed of two expressions containing variables, and their solutions are formed of expressions which are substituted to the variables so as to make the two expressions equal. For instance, a solution to the equation $x + 2 = 4$ is an expression a such that the proposition $a + 2 = 4$ is provable. Solving the equation therefore consists in giving an expression a and a demonstration of the proposition $a + 2 = 4$. In this example, when one gives the solution 2, it is not necessary to prove the proposition $2 + 2 = 4$; indeed, it suffices to compute $2 + 2$ in order to obtain the result 4. It thus seems that there exist two types of equations in mathematics: those for which a solution must come equipped with a proof of its validity, and those for which a simple computation is all the verification needed. Many of the equations encountered in high school belong to the second category. Nowadays, new, rather general solving methods have begun to flourish for such equations; however, these methods often remain less efficient than the specialized ones used in high school.

In particular, it was thanks to one such equation solving algorithm that, on October 10, 1996, William McCune's EQP program, after eight days of uninterrupted computation, proved a theorem about the equivalence of different definitions of the notion of Boolean algebra. This result may be an anecdote, but it had never been successfully proven before. This feat may not even come close to those prophesied by the pioneers of automated proof, but is nonetheless commendable.

CHURCH'S TYPE THEORY

The aforementioned methods enable one to seek proofs in simple theories, such as the theory of associativity, where all axioms are equalities. In order to demonstrate actual mathematical theorems, it was natural to try to conceive specialised automated theorem proving programs, that is, programs specific to certain theories in which the whole range of mathematics can be expressed, such as Church's type theory, or set theory.

We have seen that Church's type theory consisted mainly in one axiom, namely the beta-conversion axiom, whose form is also an equality. In 1971, Peter Andrews conceived the project of building in this axiom to the unification algorithm, much like Plotkin did with the associativity axiom. The following year, Gérard Huet carried this project through to a successful conclusion: he devised a "higher order" unification algorithm that built in the beta-conversion axiom. In other words, he transformed the beta-reduction axiom back into what it originally was, namely a computation rule. Huet suggested turning this axiom back into a computation rule at the same time Martin-Löf did – although independently and for different reasons.

Despite differences in goals and forms, the methods offered by Plotkin and Huet thus share an essential similarity: their starting points both imply transforming axioms into computation rules. This fact may help us provide a partial answer to the question that opened this chapter, namely: what ideas sped up the progress of automated proof methods since the fifties? One of the ideas crucial to that progress was the one that advocated transforming axioms into computation rules and hence distancing oneself from predicate logic and the axiomatic method. Had one preserved the axiomatic conception of mathematics, in order to prove the proposition "$2 + 2 = 4$," one would be doomed to conceive strenuous methods potentially calling upon all the axioms in mathematics, instead of simply doing the addition.

CHAPTER ELEVEN

Proof Checking

ONCE IT WAS ESTABLISHED that automated proof was not keeping all its promises, mathematicians conceived a less ambitious project, namely that of proof checking. When one uses an automated theorem proving program, one enters a proposition and the program attempts to construct a proof of the proposition. On the other hand, when one uses a proof-checking program, one enters both a proposition and a presumed proof of it and the program merely verifies the proof, checking it for correctness.

Although proof checking seems less ambitious than automated proof, it has been applied to more complex demonstrations and especially to real mathematical proofs. Thus, a large share of the first-year undergraduate mathematics syllabus has been checked by many of these programs. The second stage in this project was initiated in the nineties; it consisted in determining, with the benefit of hindsight, which parts of these proofs could be entrusted to the care of software and which ones required human intervention. The point of view of mathematicians of the nineties differed from that of the pioneers of automated proof: as we can see, the idea of a competition between man and machine had been replaced with that of cooperation.

One may wonder whether it really is useful to check mathematical proofs for correctness. The answer is yes, first, because even the most thorough mathematicians sometimes make little mistakes. For example, a proof-checking program revealed a mistake in one of Newton's demonstrations about how the motion of planets is subjected to the gravitational attraction of the sun. While this mistake is easily corrected and does not challenge Newton's theories in the slightest, it does confirm that mathematical publications often contain errors. More seriously, throughout history, many theorems have been given myriad false proofs: the axiom of parallels, Fermat's last theorem (according to which, if $n \geq 3$, there exist no positive integers x, y and z such that $x^n + y^n = z^n$) and the four-color theorem (of which more in Chapter 12) were all allegedly "solved" by crank amateurs, but also by reputable mathematicians, sometimes even by great ones. Since one can determine whether a proof is correct just by doing a simple calculation (which consists in making sure that each step of the proof is based on the application of an inference rule using only already demonstrated propositions as premise), it seems natural to entrust this calculation to a tool.

Using proof-checking programs involves writing out proofs in abundant detail, much more so than the norms of mathematical writing traditionally require. Undoubtedly tedious, this requirement nevertheless had benefits, because it encouraged greater mathematical rigor. Throughout history, the norms of mathematical writing have steadily evolved toward greater rigor and exactness, and the use of proof-checking programs constituted one more step in this long evolution: proofs had become rigorous enough for computers to be able to verify their correctness.

Had mathematics remained in the state it was in back in Newton's days, this kind of tool would probably only have taken on minor importance. In the twentieth and twenty-first centuries, however, mathematical proofs have become longer and longer, not to mention more and more complex, so that these tools are on their way to becoming indispensable – in the near or distant future, they will be necessary to check the correctness of certain types of demonstrations. For instance, whereas the demonstration of Fermat's little theorem (according to which, if p is a prime number, then p is a divisor of $a^p - a$), which was proved by Fermat himself in the seventeenth century, takes but half a page, that of his last theorem, on the other hand, which we mentioned earlier and which was proved by Andrew Wiles in 1994, covers several hundreds of pages. In this case, many mathematicians proofread the demonstration and reached the conclusion that it was correct, although they did initially spot a mistake, which Wiles managed to correct. Checking a proof's correctness without the aid of specialized programs is possible – we have seen it done in the past – but it represents huge amounts of work, so that it is doubtful whether proofreading by fellow mathematicians will continue to be a sufficient device if the length of proofs keeps on increasing. Indeed, Wiles's proof does not even rank among the longest: the proof of the classification theorem for finite simple groups, completed by Ronald Solomon in 1980, is made up of fifteen thousand pages, spread over hundreds of articles written by dozens of mathematicians. Although proof-checking programs are not yet able to deal with demonstrations of such size, they bring a ray of hope that these gargantuan demonstrations might one day be tamed.

THE AUTOMATH PROJECT

In the first proof-checking program, developed by De Bruijn in 1967 and named Automath, proofs were already built with axioms, inference rules, and computation rules – although the last merely consisted of beta-reduction and of the replacement of a defined symbol with its definition. Thus, in order to prove that $2 + 2$ equals 4, one could not content oneself with carrying out the addition: it was necessary to conduct a reasoning process. De Bruijn noted the paradox inherent in using a computer without letting it perform an addition, but he seems to have been reluctant to use more computation rules in demonstrations.

Later, it became clear that these programs would be unusable if they required the building of a reasoning process whenever a proof of the proposition $2 + 2 = 4$ was wanted. This accounts for the fact that some programs shunned set theory for a formalization of mathematics that allowed them to combine reasoning with computation, such as Martin-Löf's type theory or some of its extensions, such as the calculus of constructions. Other such programs resort to Church's type theory, but always in one of its variations where the beta-conversion axiom is replaced with a computation rule, and often enriched with other computation rules.

AFTERWORD COMPUTABILITY

The development of proof-checking programs made for the invention of new ways of using computation rules in mathematical proofs.

In Chapter 2, we have seen that some notions were given an algorithmic definition, while others were not. For example, the definition of the notion of composite number according to which a number, x, is composite when there exist two numbers x and y, greater than 1, such that x equals $y \times z$, does not immediately suggest an algorithm; many algorithms, however, enable one to decide whether a number is composite or prime. One needs only to add to the aforementioned definition the fact that the numbers y and z must be smaller than x in order to make it algorithmic – indeed, to find out whether a number x is composite, one needs only to multiply all numbers smaller than x and check whether one of these multiplications produces the result x. Yet there exist better algorithms than this one: for instance, the algorithm that consists in trying out the divisibility of the number x by all the smaller numbers!

One may thus define an algorithm, p, which determines if a number is composite – in other words, which produces the number 1 or 0 accordingly as the number it is applied to is even or odd. Using this algorithm, one may give a new definition of the notion of composite number: a number x is composite if $p(x) = 1$. And it is not hard to prove that these two definitions are equivalent, that is, that $p(x) = 1$ if and only if there exists numbers y and z such that x equals $y \times z$. There are now two ways of proving that the number 91 is composite. The first consists in giving numbers y and z of which 91 is the product, for example, 7 and 13. The second simply requires proving the proposition $p(91) = 1$ and, since this proposition is identical, after computation, to $1 = 1$ (p being an algorithm), it can be demonstrated with the axiom "for all x, $x = x$." Computation rules thus allow one to simplify and automate the construction of many proofs: those which, like "$2 + 2 = 4$," rest on a notion supplied with an algorithmic definition, but also propositions such as "91 is a composite number," which rest on notions that are not provided with algorithmic definitions and for which an algorithm was devised afterwards.

These proofs are shorter to write than traditional proofs but in order to check their correctness, one needs to carry out quite a number of computation operations.

THE CORRECTNESS OF PROGRAMS

Before proof-checking programs for general mathematics had even been conceived, computer scientists realised the necessity of conceiving proof-checking programs to verify the correctness of such and such computer program or electronic circuit. Indeed, mathematical proofs are not the only objects whose length and complexity has kept increasing in the last years of the twentieth century: the length of programs and the size of circuits underwent an even more spectacular boom. Some computer programs are composed of millions of lines, whereas this book, for instance, contains only a few thousand. Programmes and circuits reached a new level of complexity, which made older industrial objects such as the steam engine or the radio set look like remnants of a bygone era.

The only way to ensure the correctness of such a complex object is to prove it. We have already seen how one can prove the correctness of an algorithm: we gave the notion of "being composite" an abstract definition on the one hand and an algorithmic one on the other hand; then, we proved these definitions to be equivalent. Thus did we prove the algorithm to be correct with regards to the abstract definition.

The size of the correctness proofs of programs and circuits is at least proportional to that of their objects, be they programs or circuits. These proofs therefore differ from classic mathematical proofs, which come in various sizes and levels of complexity whatever the brevity of their objects. For this reason, if one constructs these new sorts of proofs "by hand," it is difficult to be sure of their correctness: a checking system proves a necessary tool. This was the reason why programs were conceived to verify the correctness proofs of programs and circuits.

In order to prove the correctness of a program or circuit, one must be able to express, among other things, the fact that a program, when applied to a certain value, produces a certain result – for example, a program using Euclid's algorithm to calculate the greatest common divisor of two numbers, when applied to numbers 90 and 21, produces the result 3. In order to define this, one may use axioms and inference rules but, since we are dealing with programs and computation, it seems more natural, here, to resort to computation rules. The first programs designed to verify the correctness proofs of programs, also known as automated theorem provers, such as the LCF program (developed by Robin Milner) or the ACL program (devised by Robert Boyer and J. Strother Moore) exploited a particular formalization of mathematical language which is neither set theory nor type theory and which comprises a programming language as sublanguage. In those languages, if you wish to demonstrate that Euclid's algorithm, when applied to numbers 90 and 21, yields the result 3, all you have to do is execute said algorithm in the programming language held within the mathematical one.

Boyer and Moore went further yet: in their language, even inference rules were replaced with computation rules. Of course, Church's theorem sets an a priori limit to this project and one cannot expect these computation rules to always terminate – nor is it the case for the rules of the ACL program. With this program, Boyer and Moore realized as much of Hilbert's program of replacing reasoning with computing as Church's theorem allowed them to.

Although they were pursuing different goals, De Bruijn and his successors on the one hand and Milner, Boyer, and Moore on the other reached a conclusion reminiscent of that reached by Martin-Löf, Plotkin, and Huet: so as to build proofs, one must rely on axioms and rules of inference, but also on computation rules.

CHAPTER TWELVE

News from the Field

THE IDEA that a proof is not constructed merely with axioms and rules of inference but also requires computation rules has come a long way: in the early seventies, this idea pervaded Martin-Löf's type theory and, since then, it also constitutes the heart of several works on the computer processing of mathematical proofs. These works study mathematical theories and proofs as objects, which they consider from the outside: in other words, these are works of logic. Mathematics, however, never evolves under the sole influence of logic. For a change to occur, something must be brought to "field" mathematics – that is, to mathematical practice.

In order to determine whether this calling in question of the axiomatic method is trivial or essential, it is important that we, too, observe it from the field. Therefore, the following chapter shall contain examples – such as the four-color theorem, Morley's theorem, and Hales' theorem – that do not deal with logic but with geometry.

THE FOUR-COLOR THEOREM

In the middle of the nineteenth century, a new mathematical problem appeared: the four-color problem. When one colors in a map, one may choose to use a different color for each region on the map. A thriftier artist may decide to use the same color twice for countries which have no common border. In 1853, this idea led Francis Guthrie to seek and find a way of coloring in a map of the counties in Great Britain using only four colors. Since sometimes four neighboring counties touch, two by two, one cannot use fewer than four colors. As a consequence, the number of colors necessary to color in this map is exactly four.

The problem of the number of colors necessary to color in a map of British counties was thus solved, but Guthrie then wondered whether this property was specific to that map, or whether it might be extended to all maps. He formulated the hypothesis that all maps could be colored in with a maximum of four colors – yet he failed to prove it. Twenty-five years later, in 1878, Alfred Kempe thought he had found the key to the problem and had managed to prove Guthrie's hypothesis to be correct, but ten years later, in 1890, Percy Heawood found a mistake in this proof. The problem was finally resolved in 1976 by Kenneth Appel and Wolfgang Haken.

Kempe's argument, albeit wrong, is worthy of attention. The usual way of coloring in a map is to start with one region, move on to the next, then color in the third one, and so on. This leads us to the point when a certain number of regions on the map have already been colored, using a maximum of four colors, and a new county awaits coloring. If, in such a situation, one can still color in the next region with one of the original four colors in such a way that no two adjacent regions receive the same color, then, by induction, that is to say step by step, one shall manage to color in any map.

When one endeavours to color in a new region, one may begin by observing the bordering, already-colored-in regions. If, by a stroke of luck, these regions do not already use all four colors, then the new region may be colored in with the color not yet received by its neighbors. On the other hand, if all four colors have already been used, one must alter the coloring pattern so as to make one of the four colors available for the new region. Kempe's proof offered a method to alter the coloring pattern of a map so as to make a color available, and it was in this proof that he made a mistake.

Although it turned out to be wrong, his attempt has the merit of inspiring the following mathematical fiction: let us imagine that Kempe's argument worked, but only for maps in which more than ten regions have already been colored in. In that case, once the first ten regions have been colored in, there would be a way of coloring in the eleventh, twelfth, thirteenth region, and so on. The four-color problem would boil down to proving that all maps presenting fewer than ten regions could be colored using only four colors.

And, since there is a finite number of ten-region maps, it would suffice to review them and color them in, one by one.

This method differs little from the quantifier elimination method which allowed us, in Chapter 4, to simplify another problem, namely solving the equation $x^3 - 2 = 0$ in the realm of integers: the quantifier elimination method showed us that it sufficed, in order to solve the problem, to look for a solution between numbers 0 and 10.

In their 1976 proof, Appel and Haken proceeded in a comparable way, although their method was far more complex. First, the property which they showed by induction was much more complicated than the simple existence of a coloring pattern; second, the finite set of maps they dealt with was a set much more difficult to define than the set of all maps containing fewer than ten regions. The general idea, however, remained similar: their proof consisted in reducing the four-color problem to a problem concerning a limited, finite set of maps. This set held 1500 maps, which one merely had to review the one after the other. Yet, had Appel and Haken tried to do this by hand, they would have died long before completion of their project. They were assisted in this task by a computer and, even so, it took 1200 hours – over a month and a half – for a result to be reached. To this day, we do not know how to solve this problem "by hand," that is, without the aid of a computer.

Using the result of these 1500 verifications, the computer could produce a proof based solely on axioms and inference rules, and print it out. But this proof would take millions of pages and no one would be capable of reading it, so the whole process would be pointless.

The specificity of this proof seems to lie in its length, which made the use of a computer necessary to build it in the first place.

Was the proof of the four-color theorem the first instance of a proof so long that its construction required a computer? The answer is both yes and no. Before 1976, computers had already been used to test the primality of large numbers, to calculate decimals of π or to provide approximate solutions for equations describing complex physical systems – such as the equation designed to calculate the temperature of each point in a crooked mechanical object. Before 1976, one had already relied on computers to prove theorems of the form "the number n is prime" where n is a number of several thousands of digits, of the form "the first thousand of decimals of the number π are 3.1415926...," or "the highest temperature in an object of shape R is 80°C" where R is the description of a complex geometrical shape – to this day, one has found no way of proving them "by hand."

However, the statement of these theorems, unlike those of the four-color theorem, is very long – be it because they contain a number made up of thousands of digits, the first thousand decimals of π, or the description of a complex geometrical shape. So that, their statements being long, their proofs are more or less bound to be long, too, and hence impossible to write by hand: indeed, a theorem's proof contains, at the very least, that theorem's statement. On the other hand, the four-color theorem did not seem fated to have a demonstration so long that only a computer could build it. Certain theorems, close to the four-color theorem in their formulation (such as the seven-color theorem, for example, which concerns maps drawn not on a plane or a sphere, like that of the earth, but on a torus) can be demonstrated on a mere couple of pages, by a traditional sort of proof. The four-color theorem thus appears to be the first theorem having both a short statement and a long proof.

SYMBOLIC COMPUTATION AND COMPUTER ALGEBRA SYSTEMS

In the early eighties, new computer programs have made the recourse to computers commonplace when proving a theorem. These are called computer algebra systems, or CAS.

We have seen that, in the eighteenth century, what with the development of calculus, new algorithms began to flourish: their peculiarity was that they were applied to functional expressions, such as $x \mapsto x \times x$, instead of numbers. It seemed natural to turn to computers in order to execute these algorithms and, for example, differentiate expressions like the one presented earlier.

These programs quickly proved useful to physicists, who were beginning to encounter trouble when faced with certain problems. For example, in the nineteenth century, Charles-Eugène Delaunay spent twenty years making approximate calculations on lunar motion, which a computer algebra system can redo in a few minutes. One such program even revealed a small mistake in

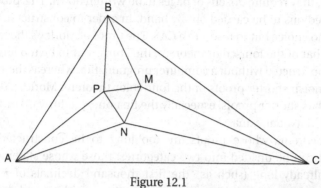

Figure 12.1

Delaunay's calculations: as he was copying an expression, he confused the fractions 1/6 and 1/16. The consequences of this error on his lunar motion studies are of little significance, which explains why it had not been detected earlier.

We have seen that Euclid's geometry was one of the few decidable theories. One of the algorithms enabling one to decide whether a geometrical proposition is provable rests on polynomial calculations, which can nowadays be done by computer algebra systems. For instance, Morley's theorem can be proved in that way. According to this theorem, given any triangle ABC, if you trisect each of its angles, you obtain straight lines which intersect in three points M, N and P, and these points always form an equilateral triangle (Figure 12.1).

There are many possible proofs to this theorem. One of them is constructed with a CAS. In this proof, one introduces six variables for the coordinates of the points A, B, and C. Then, according to these six variables, one expresses the trisectors' equations, the coordinates of the points M, N, and P, the distances MN and NP and the difference between these distances, and one obtains a functional expression which simplifies to the value 0. The distance MN is therefore equal to the distance MP. Similarly, it can be shown that the distance MN is equal to the distance NP and hence the triangle is proved to be equilateral.

At the beginning of this proof, the functional expressions are relatively short but, as it is developed, they grow larger and larger. Very soon, they require dozens of pages to be written down. This proof is too tedious to be carried out by hand. In order to construct it, one has no choice but to resort to a CAS. The proof of Morley's theorem, like that of the four-color theorem, thus turns out to be too long to be constructed without a computer program. But, whereas there are no known shorter proofs of the four-color theorem, Morley's theorem has shorter proofs, especially the first one, put forward in 1909 by M. Satyanarayana.

Theorems whose proofs are too long to be constructed by hand can be divided into two categories: those whose statements are already long (such as "the first thousand decimals of π are 3.1415926..."), and those whose statements are short and do not foreshadow the length of their proofs (such as Morley's theorem and the four-color theorem). This second category can be divided into two subcategories, the first one comprising those theorems that have both long and short proofs (such as Morley's theorem), and those that only have, as of yet, long proofs (such as the four-color theorem).

HALES'S THEOREM

Until the very end of the twentieth century, the four-color theorem was alone in the category of theorems whose statements were short and whose only known proof was long. Since it was the sole example of such a theorem, the fact that it had to be proved by a computer was not measured to its full extent. The methodological revolution at work was trivialized: many mathematicians believed this theorem would eventually be provided with a shorter proof and this would toll the knell of those scandalous, disproportionate demonstrations which were too long to be hand-written. In 1989, however, a second theorem was added to the category of theorems with short statements and long proofs, namely the proof of the nonexistence of a finite projective plane of order 10. Then, in 1995, the double bubble theorem joined their ranks, closely followed, in 1998, by Hales's theorem, also known as Kepler's conjecture. According to the latter, when you stack oranges, you can never use up more than 74% (more precisely, $\pi\sqrt{2}/6 = 0.74048...$) of the space, and so you are bound to leave at least 26% of empty space between the oranges. This theorem resolved a problem that was almost four centuries old, since it had been formulated by Kepler in 1610. Thus, the proof of the four-color theorem no longer appears to be an isolated case.

Two causes account for the length of these demonstrations: either they contain a very long case study, as does the proof of the four-color theorem, or they use long functional expressions, as does Morley's theorem – and sometimes, like the Kepler conjecture, they do both.

IS THE PROOF OF THE FOUR-COLOR THEOREM REALLY THAT LONG?

The previous theorems, whose proofs require the use of a computer, are so long that they cannot be demonstrated by hand.

However, if one chooses to construct these proofs with axioms, rules of inference, and computation rules, one rids them of all the computation tasks executed by the computers – thus, one could prove that 91 is a composite number without going over the painstaking details of the multiplication of 7 by 13. The proofs thus become much shorter to write – although, admittedly, longer to check.

UNDERSTANDING WHY

When this new type of proof appeared, in 1976, it caused a massive stir comparable to the crisis of non-Euclidean geometries or to that of constructivism. Once more, the question was posed: what exactly was acceptable in a mathematical proof? The debate no longer revolved around which axioms and rules of inference could be used; the issue was the following: when a proof was too long to be carried out by hand, was it still, in essence, a proof?

This dispute was kept under control as long as the proof of the four-color theorem remained an isolated case. It was not until the nineties and the early twenty-first century, when the number of proofs of that type began to increase, that the crisis really broke out: it had become clear that these were not marginal phenomena, but that a major change was afoot at the very heart of mathematics, and it became a matter of urgency to decide whether these proofs were to be accepted or rejected. These new proofs were reproached with two things: they were criticized for not being explicative on the one hand, and for being hard to prove correct on the other.

Let us take the example of the proof of the four-color theorem. One may argue that this proof is not explicative because, whereas it is true that all maps may be colored with four colors, there must be a reason to this fact, and there cannot exist one thousand and five hundred different reasons all miraculously agreeing on that fact. When one throws a die one thousand and five hundred times and gets a six one thousand and five hundred times one suspects a reason – the die, for example, could be loaded – one does not just assume Lady Luck is feeling generous! The very principle of the scientific method is the necessity to find a unique reason accounting for the regularity of phenomena. This reason is called an explanation. Traditionally, it is believed that a proof should produce such a reason – yet Appel and Haken's does not.

That Appel and Haken's proof is no explanation is generally agreed upon. Besides, there is a general tendency within the mathematical community to hope that the four-color theorem might one day be provided with a shorter, more explicative proof.

Is this hope ground enough, however, to reject the proof we know today? Proving a theorem by giving several arguments which all work in different cases – a "case-by-case argument" – is an old practice. For example, the proof that a square is never twice another square (see Chapter 1) proceeds by distinguishing four cases (where x and y are, in turn, even or odd). Ruling out case demonstrations from mathematics would not only invalidate the proof of the four-color theorem, but also many others: all proofs in which one distinguishes between cases where a natural number is even or odd, positive or negative, prime or composite, equal to or different from 2, for example. In a word, nearly all mathematical proofs would be rejected as well.

Of course, not all case demonstrations provoke such disapproval. Only those comprising a great number of cases draw criticism. Yet what exactly is a "great number"? Where does one draw the line between what is and what is not acceptable as a demonstration? If a proof resting on four cases is deemed acceptable, it seems difficult to reject a proof using one thousand five hundred cases without setting an arbitrary limit.

So the proof of the four-color theorem seems to be an acceptable proof, after all. And the fact that it is not explicative merely indicates that the notion of proof needs to be clearly distinguished from that of explanation.

IS THE PROOF OF THE FOUR-COLOR THEOREM CORRECT?

For all that, these new proofs still do not find favor with their critics, who reproach them with being hard to prove correct.

Whenever a new theorem is proved by a mathematician, the demonstration is proofread by other mathematicians before it is published. This system, albeit not foolproof, suffices to remove many erroneous demonstrations. Once published, the proof can be read and scanned for mistakes by just about anyone. Thus did Heawood find an error in Kempe's proof of the four-color theorem ten years after its publication.

When Appel and Haken's proof of the four-color theorem was first published, some mathematicians attempted to read and verify it. This task was much harder than the proofreading of traditional demonstrations, for they had to redo all of Appel and Haken's calculations and, in 1976, redoing a month and a half's worth of computer calculation was quite an ordeal. Only in 1995 did Neil Robertson, Daniel P. Sanders, Paul Seymour, and Robin Thomas check the proof and calculations and, even then, they were assisted in this task by a computer. Between 1976 and 1995, the correctness of the proof remained unsure – one could not rule out the possibility that Appel and Haken had made a mistake when writing their program. The notion of "reproducibility" of results – a familiar notion in physics and biology – was new to the realm of mathematics. This led some to claim that this subject was becoming an experimental science – but we will deal with the comparison between computation and experimentation later.

One possible way of eradicating any doubt concerning the correctness of this proof consisted in proving that the programs used by Appel and Haken were themselves correct. Unfortunately, this was never done. It would have been but a semisatisfactory solution, anyway, since the global proof would have been written in a strange mumbo-jumbo of traditional mathematical language on the one hand and of the language of programs used to check the program's correctness proofs on the other. Between these two languages, it would have remained doubtful whether the properties of the programs proved in one part of the demonstration were the same as those used in the other.

The difficulty of proving the correctness of proofs has been emphasized rather radically by the twelve mathematicians responsible for checking the correctness of Hales' theorem. After years of work, they declared, using a formulation new to mathematics, that they were 99 percent certain of the correctness of Hales' proof.

The best way to clear the doubts concerning the correctness of the proof of the four-color theorem was to use a proof-checking program, and this was exactly what Georges Gonthier and Benjamin Werner did in 2005: they rewrote Sanders, Seymour, Robertson, and Thomas' proof using the *Coq* program, a proof-checking program founded on the calculus of constructions – an extension of Martin-Löf's type theory. In this proof, some notions are given: first, a noncomputable definition, and then, an algorithmic one (like the notion of a composite number, in Chapter 11). These algorithmic definitions constitute an analogue to Appel and Haken's programmes. Proving the computable and noncomputable definitions to be equivalent constitutes the proof that these programs are correct. But these proofs of the correctness of programs are expressed in the same language as the rest of the theorem's proof; as a consequence, the whole demonstration is expressed in a single language. One therefore avoids the main form of error, the kind of error one feared to encounter in the previous demonstrations, namely mistakes made while writing the program.

Can this proof be wrong? Of course it can, for no verification, be it conducted by a man or a machine, is foolproof, and the *Coq* software itself could be flawed, miss errors, and accept erroneous demonstrations. One potential source of mistakes, however, has been removed: program-writing errors.

In the early twenty-first century, the proof of Hales's theorem also has been rewriten using another program called HOL. This project took several years and has only been completed in 2014.

THE SIZE OF PROOFS AND CHURCH'S THEOREM

All of the criticism leveled against these new proofs pointed to the same question: are there shorter alternatives to these proofs? For example, is there such a thing as a short proof of, say, the four-color theorem? Conservative mathematicians would observe that, in the past twenty-five centuries, ever since mathematics was born, short proofs have proved extremely efficient, and argue that, most theorems being equipped with a short proof, there is no obvious reason why the four-color theorem should not have one, too. Enthusiasts argue, on the other hand, that short proofs have been the sole tool that mathematicians could rely on until the twentieth century, and that it is pointless to compare their efficiency to that of long proofs while grounding one's argumentation on the observation of an era during which one lacked the necessary tools to build long proofs.

Should this debate remain informal? Or might mathematics itself clear up the misunderstanding? The theory of computability sheds some light on the debate about the size of proofs – even though that light is still currently dim. Indeed, one may wonder whether there might exist a link between the size of a proposition and that of its proof. Alas, the theory of computability provides this question with a negative answer – since a consequence of Church's theorem is the existence of theorems whose statements are of length n and whose shortest proof is at least of the length $1000n$, or even 2^n. Once more, let us resort to "reductio ad absurdum": let us assume that all provable propositions have a length smaller than 2^n, where n is the length of the proposition itself. In that case, there would exist an algorithm to determine whether a proposition has a proof or not, since there is but a finite quantity of texts whose size is smaller than 2^n. Simply reviewing these texts would supply us with an algorithm to decide whether the proposition has a proof smaller than 2^n, that is, whether it has a proof – which would contradict Church's theorem.

Of course, this argument says nothing of the specific case of the four-color theorem, and the few examples produced by the theory of computability of short propositions which only have long proofs are artificial ones, constructed in order to prove this theorem.

However, this shows that it would be foolish to believe that mathematics is always easy – that all that is provable can be proved in a couple of pages. Scientists always long to find simple explanations for the phenomena they study but, sometimes, the phenomena are complex, and one has to bear with this complexity.

CAN IT BE PROVED THAT A THEOREM ONLY HAS LONG PROOFS?

The question asked here is somewhat reminiscent of another question, which was raised earlier in the history of mathematics. Another consequence of Church's theorem is Gödel's famous incompleteness theorem, which, although proved in 1931 – five years earlier – can be seen as a simple consequence of Church's theorem. As in the case of the hypothesis according to which all provable propositions have a short proof, an enumeration argument suffices to show that the hypothesis according to which all propositions are either provable or refutable in set theory is contradictory. There exist propositions that are neither provable nor refutable in set theory. In 1931, when Gödel proved this theorem, the only instances of propositions neither provable nor refutable were artificial ones, constructed for the sole purpose of proving this theorem. Only in the sixties did Paul Cohen's work show that the "continuum hypothesis," an old mathematical problem which Cantor had failed to resolve in the late nineteenth century, was such an undetermined proposition.

Hence a new research avenue worth exploring: one might want to try to prove that the four-color theorem, or Hales's theorem, or any such theorem, has no short proof. Admittedly, it is unsure, as of today, how exactly to tackle this problem.

NEW SPACES AWAIT EXPLORATION

If one should try and formulate Appel and Haken's proof using only axioms and rules of inference, one would obtain a proof of several dozens of millions of pages. However, this proof only takes a few dozen or so pages when formulated with axioms, rules of inference, and computation rules. We owe it to the existence of this short proof that Appel and Haken were able to prove this theorem. From Church's theorem, it also follows that, in a theory defined by axioms, rules of inference, and computation rules, there also exist propositions that only have long proofs. For these, computation rules shall not be of much use. So it appears that there are various degrees in a proof's unattainability, since there are:

- Propositions which have short axiomatic proofs;
- Propositions which have no short axiomatic proofs, but which have short proofs provided one resorts to computation; and
- Propositions which have only long proofs, even when resorting to computation.

Until the seventies, all mathematical theorems belonged to the first category. Only in the seventies did we begin exploring the second category. The recourse to computation, however, shall not open the gates of the third category for us. Nowadays, it seems irremediably out of reach.

Before jumping to depressing conclusions, however, let us remember that, to mathematicians of the first half of the twentieth century, the gates to the second category must have seemed hopelessly closed.

CHAPTER THIRTEEN

Instruments

ASTRONOMERS used to observe the sky with the naked eye until, in the early seventeenth century, Galileo made his first telescope – or, as some historians would have it, simply pointed a spyglass skywards. Similarly, biologists originally observed living organisms with the naked eye until Anton van Leeuwenhoek started using a microscope. Thus, in the history of many sciences, there are two distinct periods, separated by the introduction of the science's first instrument.

Until the seventies, mathematics was practically the only science not to use any instruments. Unlike their lab-coated colleagues, mathematicians only required a blackboard and some chalk to make their science progress. This peculiarity can be explained by the fact that mathematical judgments are analytic: they do not require any interaction with nature and, above all, they require no measurements. Telescopes, microscopes, bubble chambers, and the like are measuring instruments or, to put it differently, instruments that extend the faculties of our senses. Therefore, it seemed natural not to need any of those when practicing mathematics.

In 1976, mathematics entered the instrumented period of its history. The instruments used by mathematicians, namely computers, are no measuring instruments; their function is not to extend the faculties of our senses, but to extend the faculties of our understanding, that is, to prolong our reasoning and computing faculties.

When an instrument is introduced in a science, a change occurs which is more quantitative than qualitative. There are many similarities between observing Jupiter's satellites with a telescope and observing the moon with the naked eye. One could imagine a human being endowed with superior eyesight who would be capable of seeing Jupiter's satellites with the naked eye, just like we are able to see the moon without the aid of instruments. Similarly, whereas the human being can only produce, by hand, demonstrations of some thousand pages, computers push back that limit by producing proofs of several million pages. However, the use of instruments can sometimes achieve a qualitative change, as well – it has transformed science in the past. For instance, when Galileo pointed his telescope at Jupiter's satellites, the observations he then made revolutionized astronomy, for the fact that the aforementioned satellites revolve around Jupiter refuted the thesis according to which all heavenly bodies revolved around the earth.

And the use of instruments is beginning to change the face of mathematics, too.

EXPERIMENTAL RESULTS IN MATHEMATICS

The fact that using an instrument gives access to new mathematical knowledge suggests that, against all odds, a judgment can be both a posteriori and analytic.

When one does a calculation with a computer or a hand calculator, one uses a physical object – the computer or hand calculator – and observes it. Such observation is necessary to establish synthetic judgments, like the one according to which the earth has a satellite.

In contrast, observation is needless and, more importantly, insufficient to establish analytical judgments such as 2 + 2 equals 4.

How can one establish that 2 + 2 equals 4 by experimentation? Let us imagine a rudimentary calculator made up of a shoe box and a few ping-pong balls. The method used to add two numbers n and p with this calculator consists in placing n balls in the box, adding p balls, and then counting the total number of balls inside the box. Let us suppose one placed two balls in the box, followed by two more. By counting the balls present in the box at the end of this operation, one reaches the result four. (As an alternative, one could use an abacus – or count on one's fingers.)

Strictly speaking, however, this experiment does not quite suffice to prove the proposition $2 + 2 = 4$. Indeed, this proposition implies that, when one places two objects in a box, followed by two more, there are then four objects in the box, whatever the objects' nature and regardless of the box's shape, temperature, and pressure. To prove this, one would have to repeat the experiment infinitely many times, replacing the ping-pong balls with tennis balls, anvils, unicorns, and so on.

On the other hand, the experiment seems to lead to the conclusion that there are four ping-pong balls in the box, right here, right now – which is a synthetic judgment – but also that $2 + 2 = 4$, meaning that, in all possible worlds, if one places two objects followed by two more in a box, in the end, the box will hold four objects – which is an analytic judgment. How can it be possible to infer an analytic judgment from an experiment?

In order to infer $2 + 2 = 4$ from this experiment, it seems necessary to establish by reasoning that if, in a single case, one puts two balls in a box, followed by another two, and obtains four, then the result will be the same in any other case – in other words, one needs to establish that, if one were to repeat the operation on Mars, with tennis balls instead of ping-pong balls, the result would still be four.

One way around this is to prove that the result of $2 + 2$ is a natural number – meaning it is one of 0, 1, 2, 3, 4, 5, and so on. The experiment seems sufficient to refute the hypothesis that $2 + 2$ is any other integer than 4. If one pieces these conclusions together ($2 + 2$ is a number, and it cannot be any other number than 4), it seems we may infer that the result is 4. As a consequence, it appears to be possible to deduce an analytic judgment from an experiment. Establishing, through an experiment, that $2 + 2 = 4$ is an analytic a posteriori judgment – of which the four-color theorem and Hales's theorem are other instances.

WIND TUNNELS AS ANALOGICAL COMPUTERS

Once the existence of analytic a posteriori judgments is established, it soon becomes obvious that such judgments are frequently used – and have been for a long time already. This discovery leads mathematicians, for example, to take a new look at numerous "experiments" carried out in natural sciences – such as the tests conducted with wind tunnels in aeronautics.

Before we delve into that matter, let us say a few words about experiments in natural sciences. As a general rule, experimentation in that field rests on hypotheses. One could, like Copernicus or Galileo, formulate the hypothesis that the trajectories of planets around the sun are circles, or, like Kepler, suggest that they are ellipses, but one could also formulate much more far-out hypotheses and surmise that these trajectories describe squares or triangles – one is free to speculate.

Hypotheses are rarely directly verifiable by experimentation or observation. But what makes astronomy a science is the fact that predictions can be made from these hypotheses. Assuming that planets describe circles or ellipses has consequences: according to whether your assumption is right or wrong, when you point a telescope at a certain point in the sky at a certain time, you either will or will not see a luminous dot. Predictions threaten hypotheses: if a prediction is not fulfilled, then the hypothesis – or at least one of the hypotheses among the ones which permitted one to make this prediction – is proved wrong. For example, the hypothesis according to which the trajectories of planets draw circles was refuted thanks to Tycho Brahe's observations and the conclusions Kepler drew from them.

The existence of verifiable predictions enabling one to eliminate some hypotheses and save others constitutes the main difference between natural sciences and speculative theories about nature. A theory about nature which does not allow one to make observable predictions is but mere speculation and there is no reason to favor this theory over any other. All the debates about the scientific or speculative nature of different theories of the psyche revolve around this idea: a scientific theory must enable one to make verifiable predictions. When one carries out an experiment in order to put a hypothesis to the test, one puts oneself in a paradoxical position, for one knows the result predicted by the theory. If one did not, the experiment would be pointless, for that would mean it did not allow one to test hypotheses.

However, when one runs a test in a wind tunnel – to measure the speed of the airflow around an airplane's wing, for instance – one does not know in advance the result predicted by the theory; indeed, the experiment is conducted in order to find it out. One does not, in doing so, attempt to test fluid mechanics – if one did, one would have to replace the plane's wing with a simpler form, so that one could determine the result predicted by the theory and compare it with the one obtained with the experiment. Therefore, these airflow tests are not experiments – not in the sense given to that word in natural sciences, anyway.

If one does not try to put the laws of fluid mechanics to the test, one may wonder what exactly it is that one seeks to do. The answer is: simply to find out the speed of the airflow around the plane. So one may suggest the hypothesis that, when one carries out a such experiment, one seeks to obtain a result via a direct measurement, with no reference whatsoever to any kind of theory. This hypothesis, albeit sometimes valid, is not satisfactory in most cases, for it supposes one tests the plane's wing in real conditions – which is rarely the case. More often than not, building a real-dimension plane wing proves too expensive and the test is conducted on a scale model.

This scale reduction is all the more striking when one conducts a test, on a lab work surface, to observe the speed of the lava flow on the slopes of a volcano. In order to make up for this change of scale, one replaces lava with a less viscous liquid. In order to pick the right level of viscosity for this liquid, the test must be conducted in the light of a theory. For instance, if the volcano's size is divided by one thousand, one needs a theory to indicate whether the liquid used to simulate lava must be a thousand times more viscous, a thousand times less viscous, or a million times less viscous.

There is another, more satisfactory explanation for the role of these tests: the initial elements of the test are a system – a plane wing, for instance – and a problem – for example, determining the speed of the airflow around the wing. For practical reasons, one cannot conduct the test in real conditions, so one resorts to a theory in order to resolve the problem. The theory enables one to reformulate the problem in a mathematical way, but the new problem produced proves too difficult to deal with by hand. So again one turns to a theory in order to conceive a scale model of the initial system, that is, another physical system whose mathematical formalization is identical or similar to the initial one. The test, when carried out on the scale model, yields the solution to the mathematical problem, hence to the initial problem.

The scale model on which the test is conducted can therefore be seen as a machine designed to solve a mathematical problem – often, it serves to resolve a mathematical problem which could be solved by computation, but would be too long to do by hand. Since the goal is to carry out a computation operation, it comes as no surprise that these tests – the ones done with wind tunnels, for instance – have, in many cases, been replaced with computer simulations. On the contrary, experiments and measurements such as those of Tycho Brahe's on the positions of planets can never be replaced by computer-assisted calculation, since they aim at gathering information about nature.

These wind tunnel tests, just like computer-aided calculation, enable one to establish analytic a posteriori judgments: they are analytic because they help solve mathematical problems, and a posteriori because they rest on an interaction with nature.

THE KNOWLEDGE NECESSARY TO THE BUILDING OF INSTRUMENTS

We have seen in Chapter 12 that the use of instruments in mathematical practice (in the proof of the four-color theorem or of Hales's theorem, for example) casts some doubt on the reliability of the results thus obtained. This doubt overlaps with another, more abstract one ensuing from the fact that these judgments are analytic a posteriori.

Knowledge achieved in mathematics differs greatly in nature from knowledge constructed in natural sciences. In mathematics, once a theorem has been proved, it is considered valid forever. For instance, two thousand and five hundred years ago, the Pythagoreans proved that a square could not be twice another square, and this demonstration still holds today.

In natural sciences, however, proofs are not the key to knowledge, hypotheses are: one lays out hypotheses, rules out those which lead to unfulfilled predictions, and saves the other ones. Yet even these hypotheses are but temporarily saved, for they are always at the mercy of a new experiment which might refute them. Therefore, in natural sciences, knowledge is hypothetical by nature – which explains why some theories of natural sciences have been abandoned, such as Ptolemy's astronomy, or the medieval "impetus" theory (according to which the motion of a stone thrown by a sling continues even when the sling has ceased its action on the stone because it was filled with impetu while inside the sling), and others were revised over the years, such as Newtonian mechanics.

Interestingly enough, when instruments are used in mathematics, mathematical knowledge is grounded in knowledge about nature – in other words, analytic knowledge is made to rest on synthetic knowledge. Is the knowledge thus gathered as reliable as the knowledge gained through demonstration alone – like the knowledge that the square of an integer cannot be twice another square?

Let us take an example. So as to do a multiplication on a hand calculator, one must first build the calculator. In order to do that, one must build semiconductors and transistors whose functioning is described in quantum physics. The correctness of the result of the multiplication done with the calculator, unlike that of a multiplication done mentally and, maybe, that of a multiplication done manually on a sheet of paper, thus depends on the correctness of quantum physics. And this theory, like all other theories of natural sciences, may one day be proved wrong. We may therefore wonder whether it really is sensible to base mathematical knowledge on revisable knowledge.

And yet, when we multiply numbers mentally and, independently, do the same multiplication on a hand calculator, and reach two different results, we are more inclined to believe that we made a calculating mistake rather than believe we just refuted the whole theory of quantum physics. In such a situation, distrusting the result displayed on the calculator's screen would be as farfetched as rejecting the result of a multiplication handwritten on a sheet of paper on the grounds that the pen's ink might be invisible and suspecting some numbers in the result to have disappeared.

Paradoxically, whereas the use of tools ought to make results less reliable, we have seen that the use of computer algebra and proof-checking programs revealed slight errors in Delaunay's calculations, as well as in Newton's, for instance. The analysis seems to indicate that the use of instruments introduces a potential source of mistakes, yet observation shows that tools often serve to correct errors. How can this paradox be unravelled?

At the origin of this paradox lies an assumption we all tend to make: whereas calculations made with instruments can be erroneous, mental calculations are foolproof for they rest on the flawless faculties of our infallible understanding. We are wary – and rightly so – of the perceptions of our senses (which could be hallucinations) and of hasty generalizations that may be made from such observations (and are always at risk of being refuted by a new, contradictory observation). When it comes to our understanding, however, we are overly confident: indeed, a philosophical tradition places the awareness of one's own understanding at the source of all philosophical reflection (for the existence of that understanding is the only thing that cannot be doubted) and misleads one into relying too much on the performance of that understanding, which, unfortunately, is as liable as one's senses to make errors, and especially calculation mistakes.

So the fact that mathematics has entered its instrumented age should not give us excessive faith in the tools used. On the contrary, it should curb the sometimes excessive trust we place in ourselves – we, too, can make mistakes.

THE COMPUTER AND THE MILLIONAIRE

This chapter dedicated to the use of instruments in mathematics would be incomplete should it not mention the way these instruments have changed the organization of mathematical work. On this point, it seems that, due to Church's thesis, which has no counterpart for the instruments used in other sciences, this change is different in mathematics than in natural sciences.

In natural sciences, one uses lenses and thermometers, telescopes and oscilloscopes, particle accelerators and bubble chambers. As for these instruments' makers, they are essentially the scientists who need them. Of course, they do not devise these tools alone on a desert island – whenever they need a bolt or a screw, they do as anyone would do: they go out and buy it. Similarly, the lenses used in astronomer telescopes, binoculars, and opera glasses are manufactured in very similar ways so that, to make a telescope, an astronomer can buy a lens from a workshop where lenses are made for sailors and opera-goers. It seems that the production of scientific instruments is not completely disconnected from industrial production. However, for the most part, scientists need tools so specific that they have no choice but to make them themselves. Their tools are often unique, hand-made specimens and, when one looks around a physics or biology lab, one encounters objects that can be sighted nowhere else.

It would seem natural for the same to apply to the tools used by mathematicians – and it did, in the early days of computer sciences: the first computers were made by mathematicians such as Turing and Von Neumann, their tools came in only one specimen and could be encountered nowhere else. Today, on the other hand, when one looks around a mathematics or computer sciences lab, one sees the same computers as in any other office. The most surprising fact about these labs is that they are absolutely unsurprising.

The fact that mathematicians use ordinary tools can be explained, in part, by Church's thesis. We have seen that, according to this thesis, all algorithms that can be done by a human being or a machine can be expressed by a set of computation rules – hence by a standard computer. Computers are universal machines that can execute all algorithms. This accounts for the fact that the very same computers are used to write mail, create images, compose music, do one's accounts, and do mathematical calculation.

This explanation is incomplete, for, although computers are all universal, different types of computers might execute different computation tasks with varying speed. In that case, different types of computers might be assigned different tasks – writing mail, creating pictures, composing music, and so on. The project of creating specialized computers, which would be faster at executing certain types of calculations, especially symbolic calculations (i.e., calculations dealing with data other than numbers, such as functional expressions, computation rules or mathematical proofs), survived until the late seventies.

This project was abandoned because these computers were made in labs, by small teams, on a small scale, while the computers for sale in the shops were mass-produced in factories, by teams composed of thousands of people. Although a lab-made computer could cut the time necessary to execute symbolic calculations by

20 percent, by the time this computer was finished, standard, industry-made computers had evolved and cut the time necessary to execute any calculation by 50 percent – thus making their hand-made rival obsolete before it was even born.

When an object's conception cost is high but the production cost of each individual item is low, it is wiser to use a mass-produced model, even if it is not optimal, than to try and conceive a specialized model. Indeed, the development effort that can be put in by hundreds of millions of potential buyers is by far greater than the one a few hundreds of specialists are capable of.

Many objects of that type can be encountered in the modern world: mobile phones were primitive when they were the prerogative of the rich; as they spread, they were perfected and their networks expanded, and they are much more sophisticated today than if they had remained the privilege of a happy few. If a millionaire had wanted to have a mobile phone and its matching network developed just for him, for all his money, he could only have afforded a system less effective than the mainstream phones and networks. As far as phones go, there are no luxury products. The same is true of the aeronautical and pharmaceutical industries.

Similarly, as computers spread, it was soon in mathematicians' best interests to use the same computers as everyone else to execute their calculations.

This denotes a new phenomenon in the way sciences use their instruments. Back in the days when instruments were so specific that they were fated to remain handmade tools, scientific activity was relatively autonomous from industrial activity: thermodynamics was necessary to the building of steam engines, but no steam engines were necessary to improve thermodynamics. Those days are gone, as is this independence.

CHAPTER FOURTEEN

The End of Axioms?

AS WE HAVE SEEN, in the seventies, the idea began to germinate that a proof is not built solely with axioms and inference rules, but also with computation rules. This idea flourished simultaneously in several fields of mathematics and computer science – in Martin-Löf's type theory, in the conception of automated theorem proving programs and of proof checking programs, but also in real-world mathematics, especially in the demonstration of the four-color theorem. As is often the case, this idea did not emerge whole, complete, and simple, for it first began to bloom in specific contexts which each influenced it in their own way: the specialists of Martin-Löf's type theory viewed it as an appendix of the theory of definitions; automated theorem proving program designers considered it a tool to make automated theorem proving methods more efficient; proof-checking program designers saw it as a means to skip small, simple steps in demonstrations; and mathematicians perceived it as a way of using computers to prove new theorems.

A few years after the emergence of a new idea, it is natural to question its impact and to seek a way of expressing it in the most general framework possible. This is what lead this author, along with Thérèse Hardin and Claude Kirchner, in the late nineties, to reformulate the idea according to which mathematical proofs are not constructed solely with axioms and inference rules but also with computation rules in the broadest possible framework, namely in predicate logic. This drove us to define an extension of predicate logic, called "deduction modulo," which is similar to predicate logic in every respect but one: in this extension, a proof is built with the three aforementioned ingredients.

Reconsidering this idea in its most basic form and rejecting the sophistication of type theory and automated theorem proving in favor of the simplicity and freshness of predicate logic enabled us to undertake both unification and classification tasks. Our initial intent was to unify the different automated theorem proving methods – Plotkin's and Huet's, in particular. Werner and I then extended this unification task to several cut elimination theorems, focusing particularly on Gentzen's and Girard's. Besides, with hindsight, several variations on type theory also turn out to differ only in the range of computation rules they allow.

The most surprising thing about this venture was that it enabled us to resuscitate a part of Hilbert's program which Church's theorem had put an end to, for calling on computation rules sometimes enables one to do without axioms altogether. For instance, the fact that the expression $0 + x$ is turned into x through computation enables one to do without the axiom according to which $0 + x = x$: as Church already noticed, the beta-reduction rule enables one to get rid of the beta-conversion axiom.

Much to our surprise did we find out that many axioms could be replaced with computation rules. Thus, we can see the dawning of a new research program. Indeed, if in predicate logic, proofs are built with axioms and rules of inference while in deduction modulo they are constructed with axioms, rules of inference, and computation rules, why not take this one step further and suppress axioms altogether so as to build proofs with rules of inference and computation rules alone?

The presence of axioms is the source of many problems, in automated theorem proving for instance, but also in cut elimination theory. Axioms have been marring mathematics ever since Hilbert's days – if not since Euclid's! Inevitably, one tends to long for a new logic in which proofs would rest on rules of inference and computation rules but be free of axioms. Hilbert aimed at creating a logic free of both axioms and rules of inference, but this program proved too ambitious and was crowned with failure. If one could manage to rid demonstrations of axioms, however, that would already constitute a major breakthrough.

Shall computation enable us to get rid of axioms once and for all? Or are we doomed, in spite of computation, to always make room for them in the mathematical edifice? Only time will tell.

Conclusion: As We Near the End of This Mathematical Voyage...

AS THIS MATHEMATICAL JOURNEY draws to its conclusion, let us cast a look at the unresolved problems we have encountered along the way and which may outline the panorama of research to come.

We have seen that the theory of computability allows one to show that, in all theories, there exist propositions both provable and short which have only long proofs, but that examples of such propositions are, as of today, mere artifices: the methods we know are too rudimentary to allow us to prove that real mathematical theorems such as the four-color theorem, Hales's theorem, or others have no short proofs. New methods must therefore be invented. Besides, the philosophical debate about the link between proof and explanation would be greatly clarified if one could state with certainty that a specific theorem has no short axiomatic proof.

Another question that remains unanswered to this day concerns the possibility of practicing mathematics without ever resorting to axioms. When axioms are compared to computation rules, they appear to be static objects: they are there, once and for all, as unchanging as they are true. Computation rules, on the contrary, enable mathematicians to do things – to shorten proofs, to create new ones, and so on. And, more importantly, thanks to the notion of confluence, computation rules interact with each other. As a consequence, every time one successfully replaces an axiom with a computation rule, there is cause to rejoice. Yet the fact that this is desirable does not always make it possible. In certain cases, we may have no other choice but to put up with axioms. The question is: in which cases, precisely?

Church's thesis has given us a glimpse of a new way of formulating natural laws. These would no longer be phrased as propositions but expressed by algorithms. Reformulating Newton's law in mechanics, or Ohm's law in electricity, should not pose much of a problem, but more recent theories such as quantum physics will require more work. However, the odds are that this work will shed new light on these theories – and maybe make them seem more concrete.

We have also seen that Church's thesis accounted for some of the astonishing efficiency of mathematics in natural sciences – but it only partly explains it. For instance, whereas it does seem to explain why a phenomenon such as gravitation is mathematizable, it does not clear up the mystery of the symmetries in particle physics. It might be interesting to determine what exactly the Church thesis explains and what it does not.

Since the four-color theorem was proved in the seventies and, more importantly, since strongly computational proofs have begun to multiply these past ten years, it has come to the attention of the mathematical community that the use of instruments allows one to push back the limits the chalk-and-blackboard technology used to impose on the size of demonstrations. No one can yet tell which of the many great open mathematical problems will be solved in the future by resorting to such instruments and which will find their solution thanks to traditional technologies. In particular, computers may very well not produce such a wealth of results in all the branches of mathematics, some of which may have a greater need for computation than others. Which ones? That is the question. The certainty that these computer-built proofs are correct will probably come only from the use of proof-checking systems. In this domain, progress is made on a daily basis will we ever reach the point where any mathematician will be able to use such systems, be he or she specialist or layman?

Last but not least, one may wonder about the impact that this return to computation will have on the form of mathematical writing. In the same way physics books now describe experiments that the reader can, to some extent, reproduce, mathematics books may in the future mention instrument-aided calculations that the reader shall be able to redo with his or her own instruments, if he or she wishes to put their correctness to the test. And readers of these books may be surprised to read in historical notes that, until the late twentieth century, mathematicians used no instruments and resolved all their problems by hand.

Biographical Landmarks

The following information is deliberately limited to themes that are related to this book's subject. These themes may be secondary in the work of the people mentioned.

People are presented in chronological order, except for our contemporaries, who are introduced in alphabetical order.

Thales of Miletus (ca. 625–ca. 546 BC) is considered the founder of geometry. He is said to have measured the height of a pyramid from that of the shadow it cast on the ground.

Anaximander of Miletus (ca. 610–ca. 546 BC) is believed to have been the first to use the concept of "unlimited," from which our concepts of infinite and of space ensue.

Pythagoras (ca. 580 – ca. 490 BC) is viewed as the father of arithmetic. One of his disciples is alleged to have proved that a square cannot be twice another square.

Plato (ca. 427–ca. 347 BC) stressed the faculty of the human conscience for gathering knowledge about nature without external aid.

Aristotle (384–322 BC) is the author of the theory of syllogism.

Euclid (ca. 325–ca. 265 BC) wrote a treatise called the *Elements* summarizing the bulk of the knowledge gathered at the time. His name is associated with the axioms of geometry, the axiomatic method, and an algorithm thanks to which it is possible to calculate the greatest common divisor of two numbers.

Archimedes (287–212 BC) determined the area of a parabolic segment and closely approximated that of a circle.

139

Ptolemy (ca. 90–ca. 168 AD) put forward a geocentric model of the universe.

Abu Ja'far Mohammad ebne Musa al-Khwarizmi (ca. 780–ca. 850 AD) wrote a book called *Al-Jabr wa-al-Muqabilah* (*Book on Integration and Equation*), which caused Indian positional writing to spread to the Arab world and then to Europe. The word "algorithm" is derived from his name.

Nicolaus Copernicus (1473–1543) suggested that, contrary to the general belief (according to which the earth stood motionless at the center of the universe), all planets, including the earth, rotated around the sun, a theory which finds precedents in antiquity. In Copernicus's theory, the trajectories of planets are circles, not ellipses.

François Viète (1540–1603) was among the first to use operations consisting in the addition or multiplication of infinite sequences of numbers. He is one of the inventors of the notion of variable.

Tycho Brahe (1546–1601) measured the positions of planets with unprecedented precision. These measures were then exploited by Kepler.

Simon Stevin (1548–1601) was among the first to use operations consisting in the addition or multiplication of infinite sequences of numbers. He is also one of the promoters of decimal notation for fractions.

Galileo (1564–1642) was among the first to use mathematics in physics. He suggested that the great book of nature was written in mathematical language. He was also one of the first to use instruments in astronomy. For these two reasons, he is regarded as the founder of modern science – which is both mathematized and experimental. We owe him the first observation of Jupiter's satellites and hence the refutation of an old belief according to which all heavenly bodies rotated around the earth. Like Copernicus and unlike Kepler, he defended the idea that the trajectories of planets were circles, not ellipses.

Johannes Kepler (1571–1630) showed that the trajectories of planets were ellipses, and not circles. He wrote a conjecture on the optimal way of stacking spheres, which was later proved by Hales (in 1998).

René Descartes (1596–1650) suggested indicating the position of a point using numbers: its coordinates. His statement, "I think, therefore I am," is an example of a synthetic a priori judgement; in this example, the mind gathers knowledge about the world from itself alone, with no external aid.

Bonavantura Cavalieri (1598–1647) is the author of a geometry of indivisibles that foreshadows calculus.

Pierre de Fermat (1601–1665) proved a theorem, Fermat's little theorem, according to which if a is an integer and p a prime number, then p is a divisor of $a^p - a$. He also formulated the hypothesis – called Fermat's last theorem – that, if $n \geq 3$ is an integer, then there exist no strictly positive integers x, y, and z such that $x^n + y^n = z^n$. This conjecture was proved in 1994 by Andrew Wiles.

Blaise Pascal (1623–1662) wrote an algorithm enabling one to calculate binomial coefficients. It is referred to as Pascal's triangle.

Anton van Leeuwenhoek (1632–1723) was among the first to use a microscope in biology.

Isaac Newton (1643–1727) was, together with Leibniz, one of the co-founders of calculus.

Gottfried Wilhem Leibniz (1646–1716) was, together with Newton, one of the co-founders of calculus. We owe him a tentative formalization of logic that prefigures Frege's.

David Hume (1711–1776) authored a principle according to which two sets hold the same number of elements if there is a one-to-one correspondence between them.

Emmanuel Kant (1724–1804) put forward the idea that mathematical judgements are synthetic a priori judgements, an idea that was fought by Frege a century later.

Carl Friedrich Gauss (1777–1855) devised an algorithm enabling one to resolve linear equation systems: this method is called Gauss elimination. He was one of the forerunners who anticipated non-Euclidian geometries.

Nicola Lobatchevsky (1792–1856) is the author of a non-Euclidean geometry in which, given a straight line and a point outside it, there can be several parallels to the line through the point.

János Bolyai (1802–1860) is the author of a non-Euclidean geometry in which, given a straight line and a point outside it, there can be several parallels to the line through the point.

Charles-Eugène Delaunay (1816–1872) studied lunar motion.

Leopold Kronecker (1823–1891) supported the thesis according to which each mathematical object must be constructed from integers in a finite number of steps. This idea makes him a precursor of constructivism.

Bernhard Riemann (1826–1866) is the author of a non-Euclidean geometry in which a straight line has no parallel.

Francis Guthrie (1831–1899) formulated the four-color problem.

Richard Dedekind (1831–1916) put forward one of the first definitions of integers, as well as one the first definitions of real numbers.

Dmitri Mendeleev (1834–1907) is the author of a periodic table of chemical elements.

Charles Sanders Peirce (1839–1914) was one of the inventors of the notion of quantifier.

Georg Cantor (1845–1918) is the founder of set theory. He proved that there were more real numbers than integers that is that there was no bijection between the set of integers and the set of real numbers. He tried to prove the "continuum hypothesis" according to which the set of all real numbers is the smallest infinite set after that of integers.

Gottlob Frege (1848–1925) wrote a logic that can be seen as the first draft of predicate logic.

Alfred Kempe (1849–1922) wrote an invalid proof of the four-color theorem.

Henri Poincaré (1854–1912) suggested that the axioms in a theory are actually definitions in disguise of the concepts used in the theory. His research on the three-body problem paved the way for the dynamical systems theory. He is considered a precursor of constructivism.

Giuseppe Peano (1858–1932) was among the first to offer axioms for the theory of natural numbers.

Frank Morley (1860–1937) was the first to formulate the theorem according to which the three points of intersection of a triangle's trisectors form an equilateral triangle.

M. Satyanarayana was the first to prove Morley's theorem.

Alfred North Whitehead (1861–1947), together with Russell, developed type theory in an important treatise called *Principia Mathematica*.

Cesar Burali Forti (1861–1931) wrote a paradox that proves Frege's logic to be contradictory.

Percy Heawood (1861–1955) wrote several contributions to the four-color theorem: he showed that Kempe's proof was wrong, demonstrated that five colors suffice to color in a map, and proved the seven-color theorem (an analog of the four-color theorem for maps that are not drawn on a plane or sphere but on a torus).

David Hilbert (1862–1943) gave predicate logic its final form. He formulated the decision problem that was to be given a negative answer by Church and Turing. Hilbert also embarked upon a program whose aim was to replace reasoning with computation but which, unfortunately, proved too ambitious. He opposed Brouwer's constructivist project.

Ernst Zermelo (1871–1953) put forward axioms for set theory. They are still used to this day.

Bertrand Russell (1872–1970) discovered a paradox simpler than Burali Forti's proving Frege's logic to be contradictory. His version of type theory corrects Frege's logic and heralds both predicate logic and set theory. Russell also developed the thesis of the universality of mathematics.

Luitzen Egbertus Jan Brouwer (1881–1966) is the founder of constructivism. Together with Heyting and Kolmogorov, he is also the father of the algorithmic interpretation of proofs.

Thoralf Skolem (1887–1963) devised an algorithm to determine the provability of all propositions in the theory of integers involving multiplication, but not addition.

Arend Heyting (1898–1980), together with Brouwer and Kolmogorov, invented the algorithmic interpretation of proofs.

Haskell Curry (1900–1982) put forward a paradox-free modification of the mathematical foundations proposed by Church. Together with De Bruijn and Howard, he renewed the algorithmic interpretation of proofs: he suggested one expressed proofs in lambda calculus.

Alfred Tarski (1902–1983) put forward an algorithm to decide the provability of all propositions in the theory of integers involving both addition and multiplication.

Andrei Kolmogorov (1903–1987), together with Brouwer and Heyting, invented the algorithmic interpretation of proofs.

Alonzo Church (1903–1995) gave a definition of the notion of "computable function," namely lambda calculus. He and Rosser proved the confluence of lambda calculus. Together with Kleene and at the same time as Turing, he demonstrated the undecidability of the halting. Again at the same time as Turing, Church proved the undecidability of provability in predicate logic. He put forward the idea that the common notion of algorithm is captured by lambda calculus and its equivalents. His project to ground mathematics in lambda calculus may have failed, but it foreshadows much future work. Church also reformulated Russell's type theory, incorporating into it some ideas taken from lambda calculus, thus creating what is nowadays known as Church's type theory.

John Von Neumann (1903–1957) built one of the very first computers.

Mojzesz Presburger (1904–1943) devised an algorithm to decide the provability of all propositions in the theory of integers involving addition but not multiplication.

Kurt Gödel (1906–1978) showed that any theory could be translated into set theory. Together with Herbrand, he gave a definition of the notion of computability, namely Herbrand-Gödel equations. He proved that constructive and nonconstructive mathematics could cohabitate within the same logic. His famous incompleteness theorem, which foreshadows Church's theorem, shows that there exist propositions that are neither provable nor refutable in type theory as well as in many other theories.

John Barkley Rosser (1907–1989) proved, together with Church, the confluence of lambda calculus. He and Kleene proved that the mathematical foundations proposed by Church were contradictory.

Jacques Herbrand (1908–1931), together with Gödel, gave a definition of the notion of computability, namely Herbrand-Gödel equations. Herbrand's theorem foreshadowed Gentzen's cut elimination theorem. Herbrand also anticipated Robinson's unification theorem.

Stephen Cole Kleene (1909–1994) gave a definition of the notion of computability, namely the notion of recursive function. Together with Church, and at the same time as Turing, he proved the undecidability of the halting problem. Together with Rosser, he showed that the mathematical foundations proposed by Church were contradictory. He was among the first to understand the connections between constructivism and computability.

Gerhard Gentzen (1909–1945) wrote a cut elimination algorithm in proofs of axiom-free predicate logic and in the theory of integers.

Alan Turing (1912–1954) gave a definition of the notion of computability founded on the notion of Turing's machine. Independently of Church and Kleene, he proved the undecidability of the halting problem and, independently of Church, the undecidability of provability in predicate logic. He put forward a thesis close to Church's. He was part of the team that built Colossus, one of the very first computers.

Robin Gandy (1919–1995) specified the physical form of Church's thesis and proved it under certain hypotheses about nature.

Nicolas Bourbaki is the collective pseudonym of a group of mathematicians (founded in 1935) who wrote an important mathematical treatise. It was in this treatise that the functional notation $\mapsto$ was first introduced. From then on, the function that to a number assigns its square was denoted $x \mapsto x \times x$.

Peter Andrews suggested incorporating the beta-conversion axiom into the unification algorithm in the aim of constructing a proof-search method for Church's type theory.

Kenneth Appel, together with Haken, proved the four-color theorem.

John Barrow glimpsed the link between Church's thesis and the effectively of mathematics in natural sciences.

Peter Bendix, together with Knuth, invented a method to transform a set of axioms of the form $t = u$ into a confluent set a computation rules.

Robert Boyer, together with Moore, invented the ACL system, founded on a formalization of mathematics that contains a programming language as sublanguage.

Nicolaas Govert De Bruijn wrote the first proof checking system, namely the Automath system. Together with Curry and Howard, he revised the algorithmic interpretation of proofs by suggesting one expressed proofs in lambda calculus.

Noam Chomsky suggested a redefinition of grammars as computation methods.

Paul Cohen proved that the continuum hypothesis was neither provable nor refutable within set theory.

Thierry Coquand, together with Huet, founded the calculus of constructions, an extension of Martin-Löf's type theory.

David Deutsch underlined the fact that the physical Church thesis expresses some properties of nature. He glimpsed the link between the thesis and the efficiency of mathematics in natural sciences.

Gilles Dowek, together with Kirchner and Hardin, invented deduction modulo as well as an automatic proof method of which several previous methods are special cases. Together with Werner, he proved a cut elimination theorem for deduction modulo of which, again, several previous theorems are particular cases.

Jean-Yves Girard, together with Tait and Martin-Löf, played a part in the development of cut elimination theory in the late sixties and early seventies. In particular, he extended the cut elimination theorem to Church's type theory.

Georges Gonthier, together with Werner, gave a demonstration of the four-color theorem using the *Coq* program.

Wolfgang Haken, together with Appel, proved the four-color theorem.

Thomas Hales proved a conjecture formulated by Kepler in 1610 and according to which, when stacking spheres, one can never use up more than 74% of the space (in fact $\pi\sqrt{2}/6$).

Thérèse Hardin, together with Kirchner and Dowek, invented deduction modulo as well as an automatic proof method, of which many previous methods are special cases.

William Alvin Howard, together with Curry and De Bruijn, revised the algorithmic interpretation of proofs by suggesting that proofs be expressed in lambda calculus.

Gérard Huet incorporated the beta-conversion axiom to the unification algorithm. Together with Thierry Coquand, he is the inventor of the calculus of constructions, which is an extension of Martin-Löf's type theory.

Claude Kirchner, together with Hardin and Dowek, invented deduction modulo as well as an automatic proof method, of which many previous methods are special cases.

Donald Knuth, together with Bendix, devised a method to turn a set of axioms of the form $t = u$ into a confluent set of computation rules.

Per Martin-Löf, together with Tait and Girard, contributed to developing the cut elimination theory in the late sixties and early seventies. He created a constructive type theory in which the notion of "equality by definition" assigned a very important role to the notion of computation.

William McCune wrote the EQP program that proved a theorem about the equivalence of different definitions of Boole's notion of algebra. This theorem had never been proved before.

Robin Milner wrote the first proof correctness checking program for programs and circuits. It was called LCF.

J. Strother Moore, together with Boyer, wrote the ACL system, a system founded on a formalization of mathematics that comprises a programming language as sublanguage.

Roger Penrose suggested that the quantum theory of gravitation might evolve, in the future, into a theory incompatible with the physical form of Church's thesis, and that the noncomputable phenomena brought to light by this new form of physics may be at work in the brain.

Gordon Plotkin incorporated the associativity axiom to the unification axiom so as to build a proof-search method for Church's type theory.

Neil Robertson, together with Sanders, Seymour, and Thomas, wrote the second proof of the four-color theorem.

Alan Robinson wrote an automated proof method called resolution that searches for demonstrations within predicate logic.

George Robinson, together with Larry Wos, wrote an automated proof method called paramodulation that searches for demonstrations within predicate logic with equality axioms.

Daniel P. Sanders, together with Robertson, Seymour, and Thomas, wrote the second proof of the four-color theorem.

Paul Seymour, together with Sanders, Robertson, and Thomas, wrote the second proof of the four-color theorem.

Ronald Solomon completed the demonstration of the classification of the finite simple groups theorem. This proof is made up of

15,000 pages, spread over hundreds of articles written by dozens of mathematicians.

William Tait, along with Martin-Löf and Girard, contributed to the development of the cut elimination theory in the late sixties and early seventies.

Robin Thomas, together with Sanders, Seymour, and Robertson, wrote the second proof of the four-color theorem.

Benjamin Werner, together with Gonthier, gave a proof of the four-color theorem using the *Coq* program. He and Dowek proved a cut elimination program for deduction modulo of which many previous theorems are special cases.

Andrew Wiles proved Fermat's theorem.

Larry Wos, together with George Robinson, proved an automated proof method called paramodulation that searches for proofs within predicate logic with equality axioms.

Bibliography

CHAPTER 1

Maurice Caveing, *Essai sur le savoir mathématique dans la Méso-potamie et l'Égypte anciennes*, Presses Universitaires de Lille, 1994.

Amy Dahan-Dalmédico, Jeanne Pfeiffer, *Une Histoire des mathématiques*, Le Seuil, 1986.

CHAPTER 2

Jean-Luc Chabert, Évelyne Barbin, Michel Guillemot, Anne Michel-Pajus, Jacques Borowczyk, Ahmed Djebbar, Jean-Clause Martzloff, *Histoires d'algorithmes, du caillou à la puce*, Belin, 1994.

Ahmed Djebbar, *L'Âge d'or des sciences arabes*, le Pommier/Cité des Sciences et de l'Industrie, 2005.

Georg Wilhem Leibniz, *La Naissance du calcul différentiel*, introduction, translation, and annotations by Marc Parmentier, preface by Michel Serres, Vrin, 1989.

CHAPTER 3

René Cori, Daniel Lascar, *Logique mathématique*, Dunod, 2003.

Gottlob Frege, *Collected Papers on Mathematics, Logic, and Philosophy*, edited by Brian McGuinness, Basil Blackwell, 1984.

Gilles Dowek, *La Logique*, Flammarion, 1995.

Paul Gochet, Pascal Gribomont, *Logique*, vol. 1, Hermés, 1990.

CHAPTER 4

Piergiorgio Odifreddi, *Classical Recursion Theory*, North-Holland, 1992.

Alan Turing, *Collected Works of A. M. Turing, Volume 4: Mathematical Logic*, edited by R. O. Gandy and C. E. M. Yates, Elsevier, 2001.

Ann Yasuhara, *Recursive Function Theory and Logic*, Academic Press, 1971.

CHAPTER 5

John D. Barrow, *Perché il Mondo è Matematico?*, Laterza, 1992.

David Deutsch, *The Fabric of Reality*, Penguin, 1997.

Robin Gandy, "Church's thesis and the principles for mechanisms," in J. Barwise, H. J. Keisler, K. Kunen, *The Kleene Symposium*, North-Holland, 1980, pp. 123–148.

Roger Penrose, *The Emperor's New Mind: Concerning Computers, Minds, and the Law of Physics*, Oxford University Press, 1989.

CHAPTER 6

Peter B. Andrews, *An Introduction to Mathematical Logic and Type Theory: To Truth through Proof*, Kluwer Academic Publishers, 2002.

Henk Barendregt, *The Lambda Calculus: Its Syntax and Semantics*, North-Holland, 1984.

Jean-Louis Krivine, *Lambda-calculus, types et models*, Ellis Horwood, 1993.

CHAPTER 7

Michael Dummett, *Elements of Intuitionism*, Oxford University Press, 2000.

Jean Largeault, *L'Intuitionnisme*, Presses Universitaires de France, 1992.

CHAPTER 8

Jean-Yves Girard, Yves Lafont, Paul Taylor, *Proofs and Types*, Cambridge University Press, 1988.

Jean-Yves Girard, "Une extension de l'interprétation de Gödel à l'analyse et son application à l'élimination des coupures dans l'analyse et la théorie des types," in Oslo University, *Proceedings of the Second Scandinavian Logic Symposium, 1970*, pp. 63–92, Studies in Logic and the Foundations of Mathematics, vol. 63, North-Holland, Amsterdam, 1971.

M. E. Szabo, *The Collected Papers of Gerard Gentzen*, North-Holland, Amsterdam, 1969.

CHAPTER 9

Thierry Coquand, Gérard Huet, "The calculus of constructions," *Information and Computation* **76**, 1988, 95–120.

Per Martin-Löf, *Intuitionistic Type Theory*, Bibliopolis, 1984.

Bengt Nordström, Kent Petersson, Jan M. Smith, *Programming in Martin-Löf's Type Theory*, Oxford University Press, 1990.

CHAPTER 10

Franz Baader, Tobias Nipkow, *Term Rewriting and All That*, Cambridge University Press, 1999.

Nachum Dershowitz, Jean-Pierre Jouannaud, "Rewrite systems," in Jan van Leeuwen, *Handbook of Theoretical Computer Science, vol. B, Formal Models and Semantics*, Elsevier and MIT Press, 1990, pp. 243–320.

Gérard Huet, "A unification algorithm for typed lambda-calculus," *Theoretical Computer Science*, 1975, 27–58.

Claude Kirchner, Hélène Kirchner, *Résolutions d'équations dans les algèbres libres et les variétés équationnelles d'algèbres*, PhD thesis, Henri Poincaré University of Nancy 1, 1982.

Donald E. Knuth, Peter B. Bendix, "Simple word problems in universal algebra," *Computational Problems in Abstract Algebra*, Pergamon Press, 1970, pp. 263–297.

G. Plotkin, "Building-in equational theories," *Machine Intelligence* **7**, 1972, 73–90.

John Alan Robinson, "A machine-oriented logic based on the resolution principle," *J. ACM* **12** (1), 1965, 23–41.

Alan Robinson, Andrei Voronkov, *Handbook of Automated Reasoning*, Elsevier, 2001.

George Robinson, Lawrence Wos, "Paramodulation and theorem-proving in first-order theories with equality," in D. Michie, R. Meltzer, *Machine Intelligence*, vol. IV, Edinburgh University Press, 1969, pp. 135–150.

CHAPTER 11

Yves Bertot, Pierre Castéran, *Interactive Theorem Proving and Program Development: Coq'art*, The Calculus of Inductive Constructions, 2004.

Keith Devlin, *Mathematics: The New Golden Age*, Penguin Books, 1988.

Jacques Fleuriot, Lawrence C. Paulson, "Proving Newton's Proposition Kepleriana using geometry and non standard analysis in Isabelle," in Xiao-Shan Gao, Dongming Wang, Lu Yang, *Automated Deduction in*

Geometry, Second International Workshop, ADG 98, Springer LNCS, 1999, pp. 47–66.

Michael Gordon, Robin Milner, Christopher Wadsworth, *A Mechanized Logic of Computation*, Lecture Notes in Computer Science, 78, Springer-Verlag, 1979.

Rob Nederpelt, Herman Geuvers, Roel de Vrijer, *Selected Papers on Automath*, North-Holland, Amsterdam, 1994.

CHAPTER 12

Kenneth Appel, Wolfgang Haken, "Every planar map is four colourable," *Illinois Journal of Mathematics* **21**, 1977, 429–567.

Samuel R. Buss, "On Gödel theorems on length of proofs, I: Number of lines and speedup for arithmetics," *Journal of Symbolic Logic* **39**, 1994, 737–756.

Claude Gomez, Bruno Salvy, Paul Zimmermann, *Calcul Formel: mode d'emploi*, Masson, 1995.

Georges Gonthier, *A computer-checked proof of the four colour theorem*, unpublished manuscript.

Thomas C. Hales, "Historical overview of the Kepler Conjecture," *Discrete Computational Geometry* **36**, 2006, 5–20.

Benjamin Werner, "La vérité et la machine," in Etienne Ghys, Jacques Istas, *Images des mathématiques*, CNRD, 2006.

CHAPTER 13

Herbert Wilf, *Mathematics: An Experimental Science*, unpublished manuscript, 2005.

CHAPTER 14

Gilles Dowek, Thérèse Hardin, Claude Kirchner, "Theorem proving modulo," *Journal of Automated Reasoning* **31**, 2003, 33–72.

Gilles Dowek, Benjamin Werner, "Proof normalisation modulo," *Journal of Symbolic Logic* **68**, 4, 2003, 1289–1316.

Printed in the United States
By Bookmasters

Printed in the United States
By Bookmasters